Oxford Studies in European Law

General Editors: Paul Craig and Gráinne de Búrca

THE BOUNDARIES OF EC COMPETITION LAW

The Boundaries of EC Competition Law

The Scope of Article 81

OKEOGHENE ODUDU

This book has been printed digitally and produced in a standard specification in order to ensure its continuing availability

OXFORD
UNIVERSITY PRESS

Great Clarendon Street, Oxford OX2 6DP

Oxford University Press is a department of the University of Oxford.
It furthers the University's objective of excellence in research, scholarship,
and education by publishing worldwide in

Oxford New York

Auckland Cape Town Dar es Salaam Hong Kong Karachi
Kuala Lumpur Madrid Melbourne Mexico City Nairobi
New Delhi Shanghai Taipei Toronto

With offices in

Argentina Austria Brazil Chile Czech Republic France Greece
Guatemala Hungary Italy Japan South Korea Poland Portugal
Singapore Switzerland Thailand Turkey Ukraine Vietnam

Oxford is a registered trade mark of Oxford University Press
in the UK and in certain other countries

Published in the United States
by Oxford University Press Inc., New York

Reprinted 2007

ISBN 978-0-19-927816-9

GENERAL EDITORS' PREFACE

Competition is central to the Community's political, economic and legal order, and Article 81 is at the core of the EC's competition provisions. The scope and meaning of Article 81 and the relationship between Article 81(1) and 81(3) have been the subject of vigorous debate, both judicially and academically, ever since the inception of the EEC. That debate shows no sign of abating, notwithstanding the broader reform to competition policy that occurred in the new millennium. We therefore welcome the addition of Okeoghene Odudu's monograph to this series, since it addresses central concerns as to the ambit of Article 81.

The author aims to reveal the intellectual order and rational structure underlying Article 81, and by doing so to provide a normative justification for its salient features. He identifies three values that can be served by competition: making the most of societal scarce resources; market integration; and economic freedom. There can be occasions in which these values can be simultaneously achieved, but they can also conflict. It is central to Odudu's thesis that when they do conflict Article 81 ranks the first of these values as superior and hence prioritizes the protection and promotion of efficiency over other concerns.

The book seeks to justify and support this thesis by considering different issues that are central to Article 81 and showing how the interpretation accorded to the particular issue serves to reinforce the thesis. This is exemplified by Odudu's treatment of the meaning of restriction of competition within Article 81, a topic on which the judiciary and academics have spent considerable time. Odudu analyses the factors that have led the Community courts to find that such a restriction exists and argues that they reveal a substantive obligation not to collude such as to cause allocative inefficiency. A restriction of competition will therefore only exist when the collusion causes allocative inefficiency, and hence restriction of competition and allocative inefficiency are regarded as synonymous. Article 81(1) is thus treated principally as the domain of allocative efficiency, while it is argued that Article 81(3) allows productive efficiency to be taken into account in the decision-making process. On this view, Article 81(3) allows a trade-off between the losses flowing from allocative inefficiency and the gains that can be forthcoming in terms of productive efficiency.

The book will be properly regarded as an important contribution to the ongoing debate in this area, and will be of interest to lawyers and non-lawyers alike.

Paul Craig
Gráinne de Búrca

PREFACE

This book addresses two problems surrounding the interpretation and application of Article 81 EC—what is competition and how does Article 81 EC ensure that competition is protected. After just short of 50 years of application, and a period of modernization, decentralization, and reflection, it is possible to understand Article 81 EC and what it seeks to achieve. Arguing that Article 81 EC imposes obligations on entities engaged in economic activity, the substantive and the justificatory rules are cast to generate obligations appropriate for such entities to perform. Actors and activities falling within the scope of Article 81 EC are subject to the substantive obligation prohibiting contrived reductions in output. Since output reduction can co-exist with cost reduction and innovation, and these latter features are desirable, cost reduction and innovation operate to justify infringement of the substantive obligation. This book seeks to show that output, cost, and innovation are the only legitimate issues in an Article 81 EC analysis; it is in this sense concerned with the boundaries of Article 81 EC.

Okeoghene Odudu
September 2005

ACKNOWLEDGEMENTS

Chapter 3 is a revised version of Okeoghene Odudu, 'The Meaning of Undertaking within 81 EC', published in (2005) 7 *Cambridge Yearbook of European Legal Studies* 209–239, and is reproduced by permission of the editors. Chapter 5 contains some material from Okeoghene Odudu, 'The Role of Specific Intent in Section 1 of the Sherman Act: A Market Power Test?', published in (2002) 25 W Comp 463–91, and such material as is reproduced is done so by permission of the publisher, Kluwer Law International.

At various stages, and in various forms, Albertina Albors-Llorens, Bill Allan, David Bailey, Oliver Black, Margaret Bloom, Paul Craig, Michele T Hearty, Josh Holmes, Stephen Weatherill, Richard Whish, and Wouter P J Wils have read, discussed, commented on, and corrected many of the ideas presented here and have been sources of much wisdom. Responsibility for all that is unwise remains mine alone.

This book develops ideas I initially conceived as a doctoral research student at Keble College, Oxford, funded by the AHRB, and supervised by Professor Paul Craig. A Kennedy Scholarship enabled some time to be spent at the ELRC at Harvard Law School. Professor Stephen Weatherill and Professor Richard Whish examined the thesis. Both gave me much to think about. Part of the re-thinking and re-working was done from room O6 of Downing College, Cambridge, whilst Fellow in law. More re-thinking and re-working has occurred since moving to my current location of L404, King's College London, School of Law, and taking up my current post as lecturer in competition law. A lifetime of thinking and re-thinking about some of the issues raised lies in store.

When I left my parents and home to commence studies as an undergraduate I had three brothers and three sisters. In the time it has taken to reach this stage, I have gained one brother, two nephews, one brother-in-law, and many friends. All may be forgiven if they forgot that they had a son, a brother, and a friend. They have not forgotten, and I would like to reassure them that they have not been forgotten. This is very much for them.

CONTENTS

TABLE OF CASES

Note on renumbering

Article 12 of the Treaty of Amsterdam (effective from 1st May 1999) provides that:

'The articles, titles and sections of the Treaty on European Union and of the Treaty establishing the European Community, as amended by the provisions of this Treaty, shall be renumbered in accordance with the table of equivalences set out in the Annex to this Treaty, which shall form an integral part thereof'.

The renumbering must be borne in mind when reading both pre-and post Amsterdam texts, and the reader must be clear as to which provision is being referred to. The discussion in the main body of this text uses new numbering. Where old numbers have been change to avoid confusion this is indicated with square brackets.

Court of Justice and Court of First Instance (Alphabetical)

Court of Justice (Numerical)

Court of First Instance (Numerical)

Commission Decisions (alphabetical)

Canada

UK Courts and Tribunals (alphabetical)

Judgments of the UK Competition Appeal Tribunal are available from http://www.catribunal.org.uk/judgments/. The versions of the judgments placed on the website may be subject to amendment. Their final versions can be found in the Competition Appeal Reports which form part of the United Kingdom Competition Law Reports (ISSN 1467 7784).

Office of Fair Trading (OFT)

Following an investigation under the Competition Act 1998, the OFT may make a decision establishing that one or more of Articles 81 and 82 and the Chapter I and Chapter II prohibitions have been infringed. In such cases, the OFT may impose penalties on the undertakings committing the infringement and give directions to bring the infringement to an end. If the OFT has made an infringement decision, it must publish its decision. If the OFT has made a decision that there are no grounds for action in respect of a particular agreement or conduct, it may publish its decision. These decisions are available from http://www.oft.gov.uk/Business/Competition + Act/Decisions/index.htm

US Cases (alphabetical)

TABLE OF LEGISLATION

EC Treaties

Regulations (chronological)

Guidelines and Notices (Chronological)

White and Green Papers (chronological

para 117 ....19, 99
para 122 139
para 125 132
para 180 17
para 185 69
para 188–192 ....68, 108
para 191 160
para 219–269 ....5, 179
para 276 19
para 293–300 21

Communication from the Commission on the Application of the Community Competition Rules to Vertical Restraints (Follow-up to the Green Paper on Vertical Restraints) Com(98)544 Final [1998] OJ C 365/3 ....3, 5, 20, 21, 100, 115, 156, 179

White Paper on Modernisation of the Rules Implementing Articles 85 and 86 of the EC Treaty Commission Programme No 99/027 [1999] OJ C 132/1 ....5, 105, 179
executive summary para 13 ....127, 177
para 4 101
para 17 172
para 49 ....3, 139
para 56 141
para 57 141
para 77 173

Proposal for A Council Regulation on the Implementation of the Rules on Competition Laid Down in Articles 81 and 82 of the Treaty and Amending Regulations (EEC) No 1017/68, (EEC) No 2988/74, (EEC) No 4056/86 and (EEC) No 3975/87 [2000] OJ C 365 E/284 141

White Paper on Reform of Regulation 17: Summary of the Observations [2001] 4 CMLR 10 ....5, 179

Commission Press Releases

Available from http://europa.eu.int/rapid/setLanguage.do?language = en

Press Release IP/01/1659 162
Press Release IP/04/411 ....5, 179

National Legislation

France

French Penal Code 1810 72

1

Introduction

I. Self-enforcing legal obligation

The function of law is to lay out the basic requirements for an ordered society.[1] The law subjects conduct to the governance of rules by which actors themselves then guide their conduct.[2] In order to guide the law must be clear.[3] A clear legal provision that is promulgated, generally applicable, and possible to comply with, becomes self-enforcing.[4] External enforcement is the exception rather than the rule because '[t]he ordinary citizen has a certain deference for law; he does not like to break the law.'[5] Wood thus expresses the view that as a minimum, those subject to competition law must 'be able to know at all times the content of the laws and the regulations to which the user is subject'.[6] Attaining this standard is vitally important, as competition law perceived as unpredictable, oppressive, or bureaucratically unfair will not command fidelity.[7]

II. The uncertain boundaries of Article 81 EC

The success of the Community in creating and applying competition law in a manner that is clear, predictable, and principled is subject to doubt.[8] For much of its history Article 81 EC has not been self-enforcing. Not because of an unwillingness on the part of those subject to comply with the law's demands, but because of an uncertainty as to what the law demands.

[1] Fuller (1969: 5–9) and Fuller (1965: 665).
[2] Fuller (1965: 657), Raz (1979: 213–14, 218, 222, 225), and Yeung (2004: 5–6).
[3] Raz (1979: 214, 222). [4] Fuller (1969). [5] Fuller (1965: 657).
[6] Wood (1993: 7). Also Goyder (1993: 3–4). [7] Goyder (1993: 5) and Slot (1993: ix).
[8] Baden Fuller (1981).

Uncertainty results because what competition is, and hence what a restriction entails, is undefined. This leaves enough 'room for argument... to ensure a comfortable old age for several battalions of lawyers.'[9] There is a view that though 'the debate about what is meant by a restriction of competition under Art. [81(1)] has been with us for 30 years' we are no 'closer to an acceptable solution to this central conundrum of competition law.'[10] The problem of what competition is requires resolution of a more fundamental question of what competition is for in the EC Treaty. Competition is certainly valued.[11] However, there is a question over whether competition is valued to promote efficiency, to achieve the Community objective of market integration, or to promote certain market freedoms desirable in a democracy.[12]

Uncertainty also arises from the way Article 81 EC is structured. Article 81(1) EC contains a prohibition; Article 81(3) EC allows the prohibition to be dis-applied if certain conditions are met. What is unclear is 'the division of labour between paragraphs (1) and (3) of Article [81].'[13] Whish asks whether 'the bifurcation of Article [81 is] the "original sin" of European competition law?'[14] It is felt that the bifurcation of Article 81 EC 'grossly

[9] Ferry (1981: 215). Also Dibadj (2004: 814–19). According to Cini and McGowan (1998: 4) the conception of competition contained in Article 81 EC may differ from the conception contained in other Community competition provisions.

[10] Panel Discussion (1998: 461).

[11] Article 3(1)(g) of Treaty Establishing the European Community: Consolidated Version [1997] OJ C 340/173 and Case C-126/97 *Eco Swiss China Time v Benetton International* [1999] ECR I-3055, para 36.

[12] On competition valued for efficiency: Green Paper on Vertical Restraints in EC Competition Policy Com (96) 721 Final [1997] 4 CMLR 519 para 54, Case T-112/99 *Métropole Télévision (M6) v Commission* [2001] ECR II 2459, para 76, Case T-313/02 *David Meca-Medina and Igor Majcen v Commission* [2004] ECR 30 September nyr, para 60–61, Levy (2003: para 1.02, 10.05(1)), Meese (2000: 472–477), and Venit (2003: 577–578). On competition valued for integration: Guidelines on Vertical Restraints [2000] OJ C291/1, para 7, Green Paper on Vertical Restraints in EC Competition Policy Com (96) 721 Final [1997] 4 CMLR 519, para 1, Albors-Llorens (2002), Fels and Edwards (1998: 63–64), Forrester QC (1998: 359–367, 371–372), Gerber (1994), Korah (1986: 91), Korah (2000: 10, 237), Neven *et al.* (1998: 37–38), Whish (2001: 19), Cini and McGowan (1998: 10–13), and Swann (2000: 129–141). On competition valued to promote freedoms: Böhm (1989: 56–57), Bundeskartellamt (2002: 4), Cini and McGowan (1998: 6), Deringer (1968: 15), Elzinga (1977: 1200–1202), Fox and Sullivan (1987: 940–944), Friedman and Friedman (1980), Furse (2004: 3–4), Gerber (2001: x–xi, 18–19, 37, 90–91, 418), Hildebrand (2002b: 153–162), Hughes (1994: 282–283), Lande (2001), Lenel (1989: 29), Levy (2003: para 2.01, 2.03, 4.03(1)), Meese (2000: 466–472), Möschel (1989: 147–148), Müller-Armack (1989: 83), Pitofsky (1979), Röpke (1960: 4–5), Scherer and Ross (1990: 18–19), and Sen (1993: 525–527).

[13] Forrester and Norall (1984: 20). Also Fox (2001: 142 n 6), Monti (2002: 1060–1061), Wesseling (2001: 359) and Whish (2000a: 78).

[14] Panel Discussion (1998: 462).

complicates, with little or no redeeming virtues, both the formulation and enforcement of specific rules'.[15] There is much debate as to what issues are appropriately addressed under Article 81(1) EC and what issues are left for consideration under Article 81(3) EC.[16] This uncertainty has been the cause of 'near anarchy'.[17]

Bifurcation produces the question of whether Article 81(1) EC as a whole, and the term restriction of competition in particular, is merely jurisdictional or whether it requires substantive assessment.[18] A classic critique of Article 81(1) EC, as enforced by the Commission, is that little or no effect on competition is required to establish an infringement.[19] Recent reforms increasingly rely on economic evidence to establish anti-competitive effect.[20] However, since Article 81(1) EC is concerned not only with the effect, but also with the object of collusion, economic effects need not be considered when the object is to restrict competition; when an economic effect must be shown is determined by the scope and content of the object category.[21] Article 81 EC remains uncertain to the extent that the meaning and function of the 'object' element is unspecified.

The uncertainty concerning what competition is, what competition is for, and how restrictions of competition are established, is further fuelled by inconsistency in how competition law is applied. Practices with similar economic consequences seem to receive dissimilar treatment under

[15] Hawk (1988: 54).

[16] Case C-309/99 *Wouters v Algemene Raad Van de Nederlandse Orde Van Advocaten (Raad Van de Balies Van de Europese Gemeenschap, Intervening)* [2002] ECR I 1577, Case T-112/99 *Métropole Télévision (M6) v Commission* [2001] ECR II 2459, Bright (1996), White Paper on Modernisation of the Rules Implementing Articles 85 and 86 of the EC Treaty Commission Programme No 99/027 [1999] OJ C 132/1, para 49, Forrester and Norall (1984: 23–24, 38–40), Green (1988: 199–201), Hawk (1988: 64–65, 69–70), Hawk (1995: 974, 977–978, 980, 982–983, 986–988), Hildebrand (1998: 188, 219, 235), Siragusa (1998), and Wesseling (2000: 88–93, 103–105).

[17] Hawk (1988: 69).

[18] Goyder (2003: 91).

[19] Green Paper on Vertical Restraints in EC Competition Policy Com (96) 721 Final [1997] 4 CMLR 519, executive summary para 37, Hawk (1995: 977–978, 982–983), Hawk (1988: 64–65, 69–70), Jones and Sufrin (2001: 171–172, 519), Siragusa (1998: 674–678), and Forrester and Norall (1984: 23–24, 38–40).

[20] Green Paper on Vertical Restraints in EC Competition Policy Com (96) 721 Final [1997] 4 CMLR 519, Communication from the Commission on the Application of the Community Competition Rules to Vertical Restraints (Follow-up to the Green Paper on Vertical Restraints) Com(98)544 Final [1998] OJ C 365/3, Commission Regulation (EC) No 2790/1999 [1999] OJ L 336/21, and Guidelines on Vertical Restraints [2000] OJ C291/1.

[21] Cases 56 and 58–64 *Établissements Consten Sàrl and Grundig-Verkaufs-GmbH v Commission* [1966] ECR 299, 342–343.

Article 81 EC.[22] Amplifying this problem is a view that the Commission and Court do not share the same view of a practice's economic consequences or method of analysis.[23] There is thus a 'complex mosaic of what is permitted and what is not permitted under Article [81].'[24]

In addition to uncertainty over the nature of the substantive obligation imposed, there is increasing uncertainty as to who must comply with the substantive obligation.[25] Article 81 EC applies to agreements and concerted practices *between undertakings* and to decisions of *associations of undertakings*. The meaning of these terms is increasingly contested.[26] At the root of the difficulty in determining whether a particular body is an undertaking rests a question of whether the entity concerned can and ought to be subject to the substantive obligation Article 81 EC seeks to impose.

There is more uncertainty: Article 81 EC only applies to *agreements* between undertakings, *decisions* of associations of undertakings, and *concerted practices*. Agreements, decisions, and concerted practices are forms of collusion; collusive behaviour is a precondition for the application of Article 81 EC.[27] The forms of collusion—agreement, decision, and concerted practice—are left undefined in the Treaty; the Court has attempted to provide definition through the cases. The result has been considerable uncertainty: Lidgard is concerned that the meaning of the collusive terms has 'not been addressed in a clear and systematic way'.[28] Wessely is of the view that '[t]he notion of collusion lacks any clear contours.'[29] Black writes that the terms have 'come to lack conceptual unity', are applied or described 'in various, obscure, confused and arguably contradictory ways',

[22] Faull and Nikpay (1999: para 7.40, 7.188–7.194), Hawk (1985: volume II, 92), Korah and Rothnie (1992: 243), Korah (1997: 206), Van den Bergh (1996), Van den Bergh and Camesasca (2001: 248–251), and Whish (1993: 544, 564, 577–578).

[23] Carlin (1996: 283), Green (1988: 199–201), Hawk (1988: 64–65, 69–70), Hawk (1995: 982–983), Korah (1993a: 19–28), Wesseling (2001: 359), Whish (2001: 548), and Whish (2000b: 889).

[24] Rose (1993: para 7–003).

[25] Lasok (2004).

[26] Case C-41/1990 *Klaus Höfner and Fritz Elser v Macrotron GmbH* [1991] ECR I 1979, Joined Case C-159/91 and C-160/91 *Christian Poucet v Assurances Générales de France (AGF) and Caisse Mutuelle Regionale du Languedoc-Roussillon (Camulrac) and Daniel Pistre v Caisse Autonome Nationale de Compensation de l'Assurance Vieillesse des Artisans (Cancava)* [1993] ECR I-637, Case C-67/96 *Albany International BV v Stichting Bedrijfspensioenfonds Textielindustrie* [1999] ECR I-5751, Joined Cases C-180/98 to C-184/98 *Pavel Pavlov v Stichting Pensioenfonds Medische Specialisten* [2000] ECR I-6451, Joined Cases C-264/01, C-306/01, C-354/01 and C-355/01 *AOK Bundesverband v Ichthyol-Gesellschaft Cordes* [2003] ECR I 2493, Note (2003), Note (2004), Belhaj and van de Dronden (2004), Drijber (2005), and Gyselen (2000).

[27] Case T-41/96 *Bayer AG v Commission* [2000] ECR II-3383, para 64. Compare with Turner (1962: 655–656).

[28] Lidgard (1997: 352).

[29] Wessely (2001: 759).

and that 'both the European Court of Justice and the domestic courts have fumbled with the concept of [concerted practice].'[30] Jakobsen and Broberg consider the law so 'unreasonably unclear' that it is 'virtually impossible to draw the line between, on the one hand, unlawful (reciprocal) agreements and, on the other, lawful unilateral acts.'[31]

In sum, there is uncertainty as to what Article 81 EC demands, who it is demanded of, and what must be done or not done to satisfy these demands.

III. Towards a self-enforcing Article 81 EC

After almost forty years in this uncertain environment, the Commission embarked on 'a fundamental review of policy' concerning certain aspects of the application of Article 81 EC.[32] Proposals for reform were made; comments on the proposals were invited; the results were summarized and subsequently published, along with a plan of action.[33] The result was Regulation 2790/1999 and the Guidelines on Vertical Restraints.[34] The momentum gathered carried us forward to the White Paper on Modernisation of the Rules Implementing Articles 81 and 82 and a Summary of the Observations.[35] On 16 December 2002 the Council adopted Regulation 1/2003 to replace Regulation 17/62 from 1 May 2004.[36] On 30 March 2004 the Commission finalized a package of notices and guidelines aimed at clarifying certain aspects of the application of Article 81 EC.[37] We are now at 'the beginning of the end of an era and style of EC competition law enforcement and the end of the beginning of a great leap forward.'[38] Nearly ten years after the process of consultation began, and with not far short of fifty years of application and reflection, it is now possible to understand Article 81 EC and what it seeks to achieve.

[30] Black (2003d: 504, 505) and Black (1992: 200). [31] Jakobsen and Broberg (2002: 139).

[32] Green Paper on Vertical Restraints in EC Competition Policy Com (96) 721 Final [1997] 4 CMLR 519, para 7, 219–269.

[33] Communication from the Commission on the Application of the Community Competition Rules to Vertical Restraints (Follow-up to the Green Paper on Vertical Restraints) Com(98)544 Final [1998] OJ C 365/3.

[34] Commission Regulation (EC) No 2790/1999 [1999] OJ L 336/21 and Guidelines on Vertical Restraints [2000] OJ C291/1.

[35] White Paper on Modernisation of the Rules Implementing Articles 85 and 86 of the EC Treaty Commission Programme No 99/027 [1999] OJ C 132/1 and White Paper on Reform of Regulation 17: Summary of the Observations [2001] 4 CMLR 10.

[36] Council Regulation (EC) No 1/2003 [2003] OJ L 1/1.

[37] Press Release IP/04/411. [38] Venit (2003: 545).

The aim of this book is to reveal the intellectual order and rational structure underlying Article 81 EC. In doing so it seeks to provide normative justification for the substantive features of Article 81 EC. After this introduction, Chapter 2 considers why competition is protected in the EC Treaty. Three constituent elements of competition's value are discussed, these being the role competition plays in making the most of society's scarce resources, the ability of competition to contribute to market integration, and the way competitive markets epitomize freedom. Whilst at times the three values can be achieved simultaneously, at other times they can conflict. It is argued that Article 81 EC ranks the ability to make the most of society's scarce resources as superior in the event of conflict and so the function of Article 81 EC is to protect and promote efficiency. It is this conception of competition that is used and shown to exist in the remainder of the book.

Who is bound by Article 81 EC is addressed in Chapter 3. Article 81 EC applies to agreements and concerted practices between *undertakings* and to decisions of associations of *undertakings*. The Court has given a functional definition to the term undertaking: the aim of the functional approach is to identify entities engaged in activities appropriately subject to the substantive obligation. The scope of the competition rules, as determined by the meaning of undertaking, must ensure that Article 81 EC is only applied to those engaged in activity in which not only does efficiency matter, but efficiency matters most. The necessity, difficulty, and consequences of the need to identify appropriate subjects for Article 81 EC is central and essential to understanding the task the Court is engaged in when determining the meaning of undertaking in Article 81 EC.

The concept of collusion plays a central role in drawing the boundary of Article 81 EC and is considered in Chapter 4, which aims to identify and develop an appropriate conception by arguing that the collusive terms capture, and ought to capture, conduct with the potential to remove uncertainty about the future conduct of others. A conception of collusion focusing on the potential to reduce uncertainty identifies conduct appropriate for scrutiny under Article 81 EC, as it is uncertainty as to the future that forces undertakings to engage in a competitive struggle. The conception of collusion advocated is broad; the breadth is unproblematic as the collusive terms can be seen to serve a jurisdictional rather than substantive function. A finding of collusion is separate from the question of whether the collusion causes harm to competition.

Chapter 5 is concerned with the central conundrum of what is meant by a restriction of competition. By examining the factors causing the Court to proclaim that restricted competition exists, it argues that the substantive obligation placed on those engaged in economic activity is not to collude to cause, or engage in collusion causing, allocative inefficiency. Finding that a restriction of competition can always and only be said to exist when collusion causes allocative inefficiency, restriction of competition as a substantive element in Article 81(1) EC and allocative inefficiency are synonymous.

Whilst Chapter 5 attempts to show that a restriction of competition within Article 81(1) EC means allocative inefficiency, it is clear that collusion causing allocative inefficiency has consequences in the productive efficiency dimension and that both efficiencies are important determinants of living standards. An exclusive concern with allocative inefficiency is undesirable and Chapter 6 argues that Article 81(3) EC provides a framework in which to assess an undertaking's claim that the allocative inefficiency consequences of their collusion are outweighed by productive efficiency gains. How Article 81(3) EC structures the productive efficiency enquiry and the trade-off between productive efficiency and allocative inefficiency is considered. In the end, it is possible to see the Article 81(3) EC elements as ensuring that the sum of allocative and productive efficiency is maximized, and that consumers benefit from the enhanced efficiency.[39]

Having argued that, as a substantive matter, Article 81 EC is concerned with efficiency, both allocative and productive, Chapter 7 seeks to show efficiency as the exclusive concern. Hard cases in Community competition law are those in which efficiency and other values cannot be pursued simultaneously. They are cases when non-efficiency values seek to triumph, cases in which reflection is had as to whether efficiency is desirable at all costs. It is acknowledged that non-efficiency goals have, in the past, been considered in Article 81 EC. However, an efficiency-only role for Article 81 EC is defended and advocated for the future as it was always inappropriate to use Article 81 EC as a Trojan horse to advance non-efficiency goals; legal certainty and justiciability are enhanced by a unitary goal, and a unitary goal maintains the intended relationship between Article 81 and other Treaty provisions. The balancing of Article 81 EC against other Treaty goals is properly external to the competition law assessment; the extent to which non-efficiency goals trump efficiency is a constitutional question external to competition law that ultimately requires resolution at the constitutional level.

[39] This is the position taken under the ECMR.

The first part of this introductory chapter set out a position of uncertainty over what Article 81 EC seeks to achieve and how it operates to achieve these aims. Chapter 8 concludes the book by considering the extent to which the arguments advanced clarify the normative boundaries of Article 81 EC. The chapters proceed by considering the questions Article 81 EC asks, the methods by which they can be answered, and what we know from the answers the questions yield. It is hoped that they provide: a method of understanding, explaining, and critiquing the law; a contribution to the evolution of what Article 81 EC is, how it is applied, and what it can achieve; and, in some small way, make the function and functioning of Article 81 EC less opaque.

2

The value of competition

I. Introduction

Competition is central to the European project. One of the EC Treaty goals is 'a system ensuring that competition in the internal market is not distorted.'[1] Article 81 EC prohibits all agreements having as their object or effect the prevention, restriction or distortion of competition within the common market. In Community law, 'Article [81] of the Treaty constitutes a fundamental provision which is essential for the accomplishment of the tasks entrusted to the Community and, in particular, for the functioning of the internal market.'[2] However, what a restriction on competition entails is not defined in the Treaty. The failure to specify what is meant by competition leads to difficulties in analysis.[3] Since competition must be defined before a restriction on competition can be identified, the primary purpose of this chapter is to specify the values justifying the protection and promotion of competition.

As noted, '[t]he debate on what the objectives of competition policy ought to be is not new.'[4] Three constituent elements of competition's value are discussed. In Section II the role competition plays in making the most of society's scarce resources is considered. Section III addresses the ability of competition to contribute to market integration. Section IV is concerned to show that competitive markets epitomize freedom. In Section V it is argued that whilst at times the three values can be achieved simultaneously, at other times they can conflict. Attention must be placed on how such conflict is resolved. The hierarchical superiority of freedom is identified in

[1] Article 3(1)(g) of Treaty Establishing the European Community: Consolidated Version [1997] OJ C 340/173.

[2] Case C-126/97 *Eco Swiss China Time v Benetton International* [1999] ECR I-3055, para 36.

[3] Dibadj (2004: 814–819).

[4] Schaub (1998: 119).

ordoliberal thought, and the European institutions have often treated integration as the superior norm. However, it is argued that Community competition law now ranks the ability to make the most of society's scarce resources as superior in the event of conflict. Section VI thus concludes that the function of Article 81 EC is to protect and promote efficiency. It is this conception of competition that is used in the remainder of the book.

II. Competition and efficiency

Article 2 EC provides that the Community shall promote *'the raising of the standard of living and quality of life'*. Economic growth is seen as essential to raise living standards.[5] Competitive markets are an effective way to achieve economic growth because of the way such markets affect the quantity, quality, and price of goods produced and consumed.[6] The effect is a consequence of two features of competitive markets—allocative and productive efficiency.[7]

A. *Allocative efficiency*

Competitive markets increase living standards by enabling society to have more: society is better off when output is maximized.[8] Whilst society is better off with increased output, if industry output is reduced, the producers can demand a higher price for each unit sold. Though firms sell fewer goods, they find it profitable to reduce output if the increased price more than compensates for lost sales.[9] It is the contrived scarcity of output rather than the increased price that is problematic; when scarcity is contrived the demand of consumers willing to pay more than the cost of production, but less than the price actually charged, goes unsatisfied.[10] This is termed deadweight loss, and consists of the value above cost that consumers would have placed on the quantity of goods no longer produced. The consequence

[5] Amramovitz (1989).

[6] Guidelines on the Application of Article 81(3) of the Treaty [2004] OJ C 101/97, para 13, 33.

[7] Marenco (1987: 430) and Prosser (2005: 17–19).

[8] The idea that economic growth increases living standards is challenged by Helleiner (1951), Weisskopf (1965), Galbraith (2004: 26–28, 73–75), and Hamilton (2004).

[9] Lerner (1934: 157).

[10] Bator (1957), Blair and Kaserman (1985: 3–22), Carlton and Perloff (2000: 69–74), Hausman and McPherson (1996: 27–30, 41–45), Motta (2004: 39–89), and Scherer and Ross (1990: 15–55).

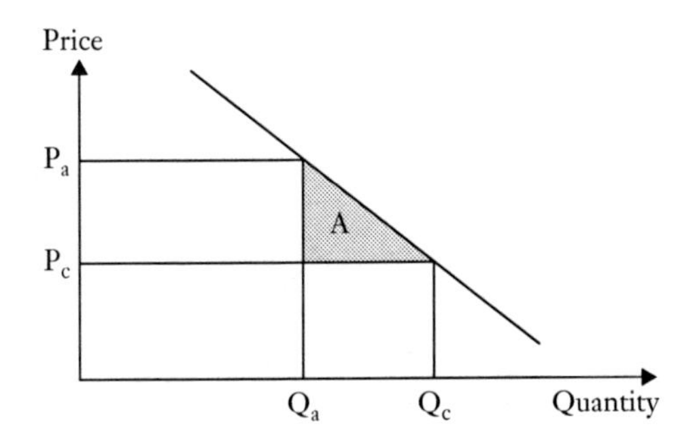

Figure 2.1 The allocative inefficiency problem.

of contrived scarcity is that society has insufficient desirable goods and pays too much for them.[11]

This allocative efficiency problem of anti-competitive conduct is represented diagrammatically in Figure 2.1.[12] The downward sloping curve represents market demand. The demand curve slopes downwards because the cheaper the price the greater the number of consumers willing and able to buy the goods. P_c and Q_c show the price and output before any anti-competitive agreements or practices take place. P_c also shows the marginal cost of production. Marginal cost is the cost of producing one additional unit of the good and (because it enables the firm to add to profit) a firm will continue producing until the marginal cost is equal to the sale price; this price includes normal profit. Anti-competitive conduct results in a price increase from P_c to P_a and a fall in output from Q_c to Q_a, with the negative consequence, deadweight loss, labelled A.

B. *Productive efficiency*

The size of the allocative efficiency problem is debatable, with some reporting the losses associated with anti-competitive conduct to be as low as 0.01 per cent of societal wealth.[13] It has been observed that since these losses were measured with a competition law regime in place, what is actually measured is the inefficiency competition law has failed to prevent.[14] Regardless of the actual magnitude of the loss, numerous commentators

[11] Begg *et al.* (1997: 138–140), Blair and Kaserman (1985: 25–37), Bork (1993: 90–115), Clarkson and Miller (1982: 107–127), Dalton and Levin (1973), Lerner (1934: 157–165), Lipsey *et al.* (1993: 286–292), and Posner (1976: 8–10).

[12] Williamson (1968a). Bork (1993: 108) considers the basic theory generally applicable to all competition law problems.

[13] Leibenstein (1966: 392–397).

[14] Posner (2001: 17–18).

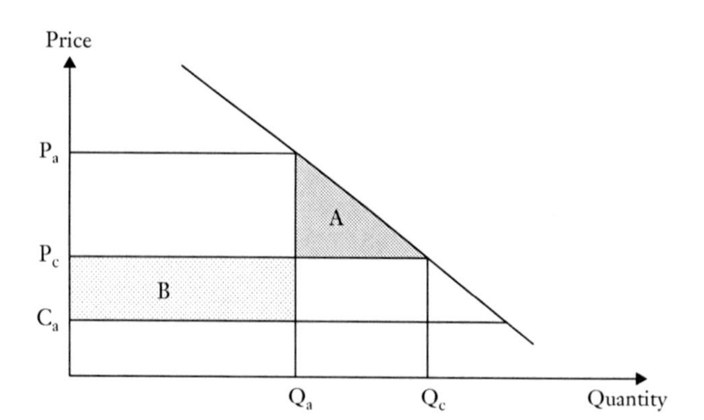

Figure 2.2 The productive efficiency trade-off.

consider that allocative efficiency is too narrow a conception of the problem of anti-competitive actions and pursue more substantial welfare consequences.[15] Rather than focus on allocative efficiency, the debate switches to whether conduct increases, is neutral, or reduces productive efficiency, since this is thought to have a greater impact on living standards.[16]

Some argue that firms with the ability to profitably contrive scarcity of output do not face pressure to minimize costs and so are productively inefficient.[17] Others argue that the ability to profitably contrive scarcity provides an incentive for firms to innovate, creating new products, services, methods, markets, and organizational formats. Whilst success confers market power, it also destroys market power that existed prior to the innovation. The ability to contrive scarcity profitably is transferred from one firm to another in the process of 'creative destruction'.

Williamson proposes a scheme, reflected in Figure 2.2, that allows us to assess the conflicting views of productive efficiency consequences of firms being able to profitably contrive scarcity of output on a case by case basis; further, productive efficiency gains that exist may be used to offset allocative efficiency losses.[18]

Productive efficiency gains are represented by the marginal cost curve falling from P_c to C_a. These cost savings are made by using fewer resources to produce Q_c than was previously the case; the freed resources can be

[15] Leibenstein (1966), Neven (1998: 113–114), Porter (2001: 922–924), Posner (2001: 13–17), Posner (1975), Williamson (1969b) and Williamson (1977: 706–709).

[16] Kendrick (1977: 1).

[17] Motta (2004: 46–49) and Hildebrand (1998: 144).

[18] Williamson (1968a: 21–23), Williamson (1968b: 1372–1373), Williamson (1969a), DePrano and Nudent (1969), Hylton (2003: 311–317, 330–332), Hildebrand (1998: 144), and Arthur (1994: 15–19).

employed elsewhere. If the disadvantage of anti-competitive conduct (area A) is less than the advantages from the same conduct (area B) society is better off as a whole. Competition law should thus consider the net welfare consequence of the ability to profitably contrive scarcity of output.[19]

III. Competition and market integration

In addition to efficiency, competition law can be used to achieve the Community objective of market integration.[20] A driver for market integration is the realization that firms in isolated markets, which concentrate solely on satisfying the demand of their home market, cannot exploit efficiency advantages because the minimum efficient scale may greatly exceed national demand.[21] In a larger, integrated market, firms have the ability and incentive to specialize, improve their technical efficiency, and exploit economies of scale and other cost advantages.[22] Once national firms have specialized, firms from all countries should then be free to trade their goods without impediment; Member States are prevented from implementing protectionist policies that allow domestic firms, unilaterally or collectively, to increase prices above the competitive level.[23] However, once state barriers to trade are removed, there is a risk that they will be replaced by private barriers to trade.[24] Competition law aids integration by preventing private firms from, unilaterally or collectively, creating impediments to trade that would enable them to set prices higher than would prevail under competitive market conditions.[25] It is thus written that '[t]he main reason

[19] Compare with Guidelines on the Assessment of Horizontal Mergers under the Council Regulation on the Control of Concentrations between Undertakings [2004] OJ C 31/5, para 8 and Panel Discussion (1998: 482).

[20] Guidelines on Vertical Restraints [2000] OJ C 291/1, para 7, Green Paper on Vertical Restraints in EC Competition Policy Com (96) 721 Final [1997] 4 CMLR 519, para 1, Albors-Llorens (2002), Fels and Edwards (1998: 63–64), Forrester QC (1998: 359–367, 371–372), Gerber (1994), Korah (1986: 91), Korah (2000: 10, 237), Neven *et al.* (1998: 37–38), Whish (2001: 19), Cini and McGowan (1998: 10–13), and Swann (2000: 129–141).

[21] Scitovsky (1956: 72–74) and Ross (2004: 246–247).

[22] Barnard (2004: 3–6), Swann (2000: 1, 99–126), Molle (2001: 14, 95–96, 102, 115, 128), Economist (2000f), Scitovsky (1956: 78–81, 88–91), and Wesseling (2000: 10–11). Contra Carchedi (2001: 36–44).

[23] Molle (2001: 333), Barnard (2004: 6–8), Cini and McGowan (1998: 11), and Organisation for Economic Co-operation and Development (1999: 4).

[24] Cini and McGowan (1998: 11, 32–33) and Monti (2002: 1062–1063).

[25] Khemani (1998: 193–198), Organisation for Economic Co-operation and Development (1999: 5), Corden (1972), Green Paper on Vertical Restraints in EC Competition Policy Com (96) 721 Final [1997] 4 CMLR 519, para 70–84, Grewlich (2001: 369–373), and Cruz (2002: 86).

behind the Community antitrust rules appeared to be the concern that the obstacles to interstate trade, banned by the Treaties, might be resurrected... by enterprises of the various Member States which would divide markets among themselves along state lines.'[26]

IV. Competition and freedoms

In addition to efficiency and integration, competition can also be used to promote and enhance freedom.[27] Certain freedoms are inherent in competitive markets: individuals are free to participate as producers and free to decide what to produce.[28] In this way, individuality is expressed. The freedom to participate in the marketplace is taken by its strongest advocates as a right, and described as 'a constituent part of private autonomy'.[29] Though the historical relationship between restraint of trade and competition law remains ambiguous, the freedom to participate is also reflected in the law on restraint of trade, a common law doctrine protecting the 'freedom to make choices such as where to work, what to produce, and what to do with the profits.'[30] In addition to the freedom to produce, competitive markets also provide a range of goods and services, allowing us to express our individuality by choosing what to consume.[31]

Competitive markets possess a mystical ability to coordinate the freedom to produce with the freedom to consume, as eloquently described by Knight:

> One of the most conspicuous features of organization through exchange and free enterprise... is the absence of conscious design or control. It is a social order, and

[26] Marenco (1987: 429–430). Also Ellis (1963: 248–265) and Guidelines on Vertical Restraints [2000] OJ C 291/1, para 7.

[27] Sen (1993: 525–527), Friedman and Friedman (1980), Pitofsky (1979), Meese (2000: 466–472), Scherer and Ross (1990: 18–19), Fox and Sullivan (1987: 940, 942, 944), Furse (2004: 3–4), Elzinga (1977: 1200–1202), Cini and McGowan (1998: 6). Contra Galbraith (2004: 18–26) and Hamilton (2004: 62–97, 103–116).

[28] Deringer (1968: 14) and Prosser (2005: 19–20).

[29] Böhm (1989: 56–57). Also Röpke (1960: 4–5), Gerber (2001: x-xi, 18–19, 37, 90–91, 418), Hildebrand (2002b: 153–162), and Bundeskartellamt (2002: 4).

[30] Gerber (2001: 19). Also Heydon (1999: 43–44, 51), Waelbroeck and Frignani (1999: 167–171), Letwin (1981: 18–52), Mason (1937: 39), Amato (1997: 7–19), Bork (1993: 20), and Gerber (2001: 208).

[31] Deringer (1968: 15), Lenel (1989: 29), Müller-Armack (1989: 83), Möschel (1989: 147–148), Fox and Sullivan (1987: 942–944), Hughes (1994: 282–283), Lande (2001), and Levy (2003: para 2.01, 2.03, 4.03(1)).

one of unfathomable complexity, yet constructed and operated without social planning or direction, through selfish individual thought and motivation alone. No one ever worked out a plan for such a system, or willed its existence; there is no plan anywhere, either on paper or in anybody's mind, and no one directs its operations. Yet in a fairly tolerable way, 'it works', and grows and changes... Innumerable conflicts of interest are constantly resolved, and the bulk of the working population kept generally occupied, each person ministering to the wants of an unknown multitude and having his own wants satisfied by another equally vast and unknown—not perfectly indeed, but tolerable on the whole, and vastly better than each could satisfy his wants by working directly for himself.[32]

The coordination occurs by using price as an 'instrument for inducing consumers to reveal their preferences and producers to cater to these allocations.'[33] The coordinated outcome reflects revealed preferences if alternatives are properly priced: the competitive process prices properly because the price is set by the forces of supply and demand.[34]

Anti-competitive conduct distorts the ability of the competitive process to reflect our expressions of freedom so that competition's protection is justified by the inherent freedoms and the ability to coordinate, without distortion, those freedoms. The process of competition has intrinsic value, and this view has heavily influenced the Community conception of competition law.[35] What is often spoken of as ordoliberalism is an attempt to preserve the freedom of the individual in the industrialized economy by constitutionalizing the competitive process and charging the state with protecting and maintaining a competitive order.[36] The belief is held that '[t]he extent of competition in the organization of industry will help to determine whether we remain a community long capable of social

[32] Knight (1951: 31–32). Also Clark (1961: 66–72), Hayek (1980: 94–96, 106), Barry (1989: 114), Böhm (1989: 53), Eucken (1989: 27–28), Gerber (2001: 20), Hayek (1980: 99–101), Müller-Armack (1989: 83), Nicholls (1994: 193), Petersmann (1999: 147–148), and Zweig (1980: 21).

[33] Bator, (1960: 77).

[34] Gerber (2001: 110), Lipsey *et al.* (1993: 220–237), and Zweig (1980: 21).

[35] Guidelines on the Application of Article 81(3) of the Treaty [2004] OJ C 101/97, para 105, Blaug (2001), Faull and Nikpay (1999: para 2.75–2.78), Fox and Sullivan (1987: 936), Gerber (2001: 1, 16, 34, 37, 41, 164, 245–246, 261–265, 271, 309, 331–334), Gyselen (2002: 184), Hildebrand (1998: 1–5), Hildebrand (2002a: 3), Hughes (1994: 297–306), Jebsen and Stevens (1996: 449–453), Korah (1993b: 162), Maher (2000: 162–165), van Miert (1998), Peacock and Willgerodt (1989b: 10), Petersmann (1999: 152–154), and Waelbroeck and Frignani (1999: 156).

[36] Müller-Armack (1989: 82), Barry (1989: 114), Böhm (1989: 50–55), Eucken (1989: 34–37), Gerber (2001: 1, 16, 34, 245–246, 250–251), Lutz (1989: 153), Lenel (1989: 21), Möschel (1989: 149), Nicholls (1994: 154, 185–189), Oliver Jr (1960), Röpke (1960: 4–5), and Zweig (1980: 17–19).

and political democracy.'[37] Like ordoliberals, Fox and Sullivan consider that competition law should 'protect not an out-come, but a process—competition. Antitrust laws set fair rules of the game.'[38] Hostility to anti-competitive conduct is not founded on its economic impact; hostility to anti-competitive conduct is based on the view that such conduct is incompatible with democracy.[39] By protecting the competitive order, the state protects the freedom of self-responsible individuals to function in the market.[40]

V. The hierarchy of values

Whilst competition law can protect and promote efficiency, integration, and freedom, these values may at times conflict.[41] A choice must then be made as to whether integration is desired at the expense of freedom and efficiency, freedom desired at the expense of efficiency and integration, or efficiency desired even at the expense of integration and freedom.[42]

Early on, Deringer wrote that 'the protection of competition against restrictions is intended to safeguard competition as an institution or to guarantee the freedom of the individual.'[43] The competitive process does not always produce efficient outcomes.[44] From an ordoliberal perspective, the competitive process should be protected, even when competition is inefficient, because freedom is the ultimate goal.[45] Röpke warns against a time when 'personal comforts are rated more highly than freedom.'[46] Arguing that we should look beyond supply and demand, he writes that 'the market economy is not everything. It must find its place within a higher order of things which is not ruled by supply and demand, free process, and competition.'[47] There is thus 'a bias in favour of the market for its expression of freedom, even if it were demonstrably less "efficient" '.[48] As

[37] Rostow (1947: 570). [38] Fox and Sullivan (1987: 959).

[39] Harding and Joshua (2003: 39–51).

[40] Böhm (1989: 55–58), Müller-Armack (1989: 84), and Zweig (1980: 24–26).

[41] Gerber (2001: x), Cini and McGowan (1998: 2–5), and Snyder (1990: 72–78).

[42] Gyselen (2002: 184) argues the values are complementary rather than conflicting. Jacobs (1995) argues that a choice between values cannot be avoided.

[43] Deringer (1968: 14). Also Waelbroeck and Frignani (1999: 168–171).

[44] Turner (1969: 1208–1209), Areeda and Kaplow (1997: 27–35), and Clark (1961: 65).

[45] Eucken (1989: 34–37), Peacock and Willgerodt (1989a: 6), Gerber (2001: 1, 16, 37), Maher (2000: 164), Monti (2002: 1059–1062), Hildebrand (2002b: 153–162), and Van den Bergh and Camesasca (2001: 136–137). Compare with Case COMP/C-3/37.792 Microsoft, para 969.

[46] Röpke (1960: 17). [47] Röpke (1960: 6). [48] Peacock and Willgerodt (1989a: 7).

Möschel writes, '[t]he actual goal of the competition policy of Ordoliberalism lies in the value of individual economic freedom of action as a value in itself... Economic efficiency... is but an indirect and derived goal.'[49] The high ranking of freedom is not uniquely ordoliberal. In *United States v Topco*, Justice Marshall opined that:

Antitrust laws in general... are the Magna Carta of free enterprise. They are as important to the preservation of economic freedom and our free-enterprise system as the Bill of Rights is to the protection of our fundamental personal freedoms. And the freedom guaranteed each and every business... is the freedom to compete—to assert with vigor, imagination, devotion, and ingenuity whatever economic muscle it can muster.[50]

According instrumental rather than intrinsic value to the competitive process would enable the competitive process to be restricted in order to achieve an outcome more preferable than that provided by the competitive process.[51] As Asch writes, since our 'ultimate concern is with policies that... will yield beneficial results... it may be argued that the presence of such results is the most sensible definition of competition.'[52] Whilst Community competition law is influenced by ordoliberalism, it is not an implementation of ordoliberalism.[53] Ordoliberalism and Article 81 EC increasingly part company when the freedoms inherent in competitive markets conflict with efficiency. The European Commission has signalled its intention to take the outcome as central, writing that:

the interest of the consumer is at the heart of competition policy. Effective competition is the best guarantee for consumers to be able to buy good quality products

[49] Möschel (1989: 146). Also Röpke (1960: 5–6) and Clark (1961: 85–86).

[50] *United States v Topco Associates Inc* 405 US 596, 610 (1972).

[51] Bork (1993: 134–135) and Kobayashi (1998: 82). The historical shift in conception of competition from process to outcome is charted by Stigler (1957). The implications of the shift are considered by Hertz (2002: 80–113).

[52] Asch (1970: 164). Also Hay (1999: 37), Bork (1993: 58–61), Bork (1967a: 252), Evans (2001), Hildebrand (1998: 137–138), McGee (1971: 13–23), Posner (1981b: 20–21), Posner (1977: 15) and Posner (1976: 22). That efficiency trumps other goals in US antitrust is challenged by Kovacic (1990: 1459–1464).

[53] Peacock and Willgerodt (1989b: 10), Oswalt-Eucken (1994: 42–43), Schaub (1998: 123–125), Amato (1997: 13) and Gerber (2001: 361–362), and Hildebrand (2002a: 16). Compare with Green Paper on Vertical Restraints in EC Competition Policy Com (96) 721 Final [1997] 4 CMLR 519, para 180, Guidelines on the Application of Article 81(3) of the Treaty [2004] OJ C 101/97, para 105, Boscheck (2000: 24–25), Cruz (2002: 1), Hawk (1988: 69–70), Hawk (1995: 973, 978–979), Monti (2002), Petersmann (1999: 151–154), Riley (1998: 483–484), Siragusa (1998: 653–654), and Wesseling (2000: 27, 31).

at the lowest possible prices. Whenever in this green paper the introduction or protection of effective competition is mentioned, the protection of the consumer's interest *by ensuring low prices is implied*.[54]

The CFI has signalled the hierarchical superiority of the outcome and the instrumental value of the process in *Métropole*, pointing to a:

trend in the case-law according to which it is not necessary to hold, wholly abstractly and without drawing any distinction, that any agreement restricting the *freedom of action* of one or more of the parties is necessarily caught by the prohibition laid down in Article [81(1)] of the Treaty.[55]

That the process (in which freedom is inherent) is of instrumental value is accepted in *Gottrup-Klim*. The case involved DLG, a co-operative society established to provide low cost farm goods to its members. Local associations who sold, purchased, or produced farm supplies could join as 'B' members and to a limited extent were entitled to take part in DLG's management. In 1975, DLG's 'B' members formed LAG and, dissatisfied with the price DLG charged, started importing fertilizers themselves through LAG. DLG amended its statutes so that 'B' members could not hold membership of a competing association and expelled thirty-seven 'B' members who refused to comply with the amendments. 'B' members questioned whether prohibiting them from participating in competing ventures restricted competition.

As amended, the DLG statutes restricted free conduct; LAG argued that the statutes were 'a significant and arbitrary limitation of the members' *freedom of commercial activity*'. LAG wanted to protect '*freedom of commercial action*', '*freedom . . . to operate independently in the market*'.[56] LAG recognized the distinction between the outcome of free competition and free conduct, arguing that the fetter on free conduct was relevant because it 'would further *reduce competition* between DLG and third parties'.[57] The position advanced was that restricting conduct or commercial freedom *necessarily* entails a

[54] Green Paper on Vertical Restraints in EC Competition Policy Com (96) 721 Final [1997] 4 CMLR 519, para 54, emphasis added. Compare with Levy (2003: para 1.02, 10.05(1)).

[55] Case T-112/99 *Métropole Télévision (M6) v Commission* [2001] ECR II-2459, para 76, emphasis added. Also Case T-313/02 *David Meca-Medina and Igor Majcen v Commission* [2004] ECR 30 September nyr, para 60–61 and Venit (2003: 577–578). Compare with Meese (2000: 472–477).

[56] Case C-250/92 *Gottrup-Klim Grovvareforening and Others v Dansk Landbrugs Grovvareselskab Amba (DLG)* [1994] ECR I-5641, para 13, emphasis added.

[57] Case C-250/92 *Gottrup-Klim Grovvareforening and Others v Dansk Landbrugs Grovvareselskab Amba (DLG)* [1994] ECR I-5641, para 13, emphasis added.

restriction on the competitive outcome.[58] Advocate General Tesauro rejected this idea, saying that a restriction on freedom is not synonymous with and does not *necessarily* entail a restriction on the competitive outcome. This is particularly so when, without the restriction on free conduct, consumers 'would have to tolerate prices which on average would be higher'.[59]

Whilst efficiency trumps freedom in Article 81 EC, integration must also be placed in the hierarchy. The Commission has expressed the view that Article 81 EC promotes integration independently of efficiency consequences.[60] Korah writes that '[c]ompetition in the common market is also intended to further market integration, an overriding aim considered by some to be more important than efficiency.'[61] Whish notes that '[u]nification of the single market is an obsession of the Community authorities . . . faced with a conflict between the narrow interests of a particular firm and the broader problem of integrating the market, the tendency will be to subordinate the former for the latter.'[62] The Commission has thus condemned agreements as restricting competition even though finding that the practice *did not* lead 'to a sales price higher' and in fact enabled the participants to lower 'the sales price by reducing his margin of profit'.[63] Such decisions are based on the overriding importance of the market integration objective.[64]

A case can be made that the rationale for integration is efficiency.[65] It may then be that market integration is used as a proxy for efficiency and that integration is valued because it normally increases efficiency rather than

[58] Case C-67/96 *Albany International BV v Stichting Bedrijfspensioenfonds Textielindustrie* [1999] ECR I-5751, AG Opinion para 252 and Wesseling (2000: 90–93).

[59] Case C-250/92 *Gottrup-Klim Grovvareforening and Others v Dansk Landbrugs Grovvareselskab Amba (DLG)* [1994] ECR I-5641, AG Opinion para 18. Compare with Gyselen (2002: 182) and Kolasky (2002). The Court rejected the claim on the basis that the restrictions were ancillary; judgment, para 28–45.

[60] Green Paper on Vertical Restraints in EC Competition Policy Com (96) 721 Final [1997] 4 CMLR 519, para 276, executive summary para 12, Guidelines on Vertical Restraints [2000] OJ C291/1, para 7, Whish (2000b: 911–917) and Cruz (2002: 99–100).

[61] Korah (2000: 10, also 237). Also Korah (1986: 91), Neven *et al.* (1998: 37–38), Cruz (2002: 96–103), Weatherill (1989: 68–73) and Schaub (1998: 126–127).

[62] Whish (2001: 19). Also XXVIIIth Report on Competition Policy 1998 SEC (99) 743 Final, para 2–3.

[63] Dru/Blondel [1965] JO 2194/65 [1965] CMLR 180, 183. Also Hummel/Isbecque [1965] JO 2581/65 [1965] CMLR 242.

[64] Green Paper on Vertical Restraints in EC Competition Policy Com (96) 721 Final [1997] 4 CMLR 519, para 117, Wesseling (2000: 80–82, 85–88), Jones and Sufrin (2001: 519), Mastromanolis (1995: 561–565, 573–588, 607–608), Korah and O'Sullivan (2002: 21–23), Faull and Nikpay (1999: para 7.07–7.10, 7.95–7.100), Siragusa (1998: 656–658), Boscheck (2000: 24–25), and Cosma and Whish (2003: 41–42).

[65] See Section III above and Van den Bergh (1996: 76–80).

independently of, and potentially conflicting with, efficiency.[66] As Blair and Kaserman write, 'when market division works well, there is no competition at all . . . market division is even worse than price fixing and certainly deserves equally harsh treatment in antitrust terms'.[67] However, that the tool of efficiency and the goal of integration can conflict is not considered, so Gerber writes:

> there was little reason to distinguish between [efficiency and integration]; they were related and mutually reinforcing. To the extent that competition law eliminated obstacles to the flow of goods, services, and capital across European borders, for example, it served the cause of unifying the market while simultaneously benefiting consumers by increasing the number of actual and potential competitors in European markets.[68]

Since conflict is not contemplated, the choice of primacy goals is not consciously made. On this view, the pursuit of integration at the expense of efficiency undermines the *raison d'être* of integration.[69] The conflict is increasingly recognized. Moreover, the conflict is increasingly resolved in favour of efficiency. For example, when analysing vertical agreements the Commission now focuses less on market integration and more on the efficiency, recognizing that such agreements cannot have a negative impact on the market unless the parties possess market power.[70] This philosophy is reflected in Regulation 2790/1999 and the Guidelines on Vertical Restraints.[71] It is increasingly seen as the function of Article 81 EC to address market power sub-dominance.[72] The Commission's 2001

[66] Guidelines on the Application of Article 81(3) of the Treaty [2004] OJ C 101/97, para 13, Guidelines on Vertical Restraints [2000] OJ C 291/1, para 7, Schaub (1998: 126–127), Wolf (1998: 130–131), and Blair and Kaserman (1985: 166–169).

[67] Blair and Kaserman (1985: 169).

[68] Gerber (1994: 103, citations omitted, also 119–120) and Albors-Llorens (2002: 312–323).

[69] Monti (2002: 1062–1063, 1069), Neven *et al.* (1998: 20), Van den Bergh (1996: 75), and Mastromanolis (1995: 614–622).

[70] Easterbrook (1984: 19–23), Easterbrook (1987: 312), Landes and Posner (1981: 937–938), Nicolaides (2000: 10), Posner (1981b: 16), and Van den Bergh and Camesasca (2001: 193–195, 247).

[71] Green Paper on Vertical Restraints in EC Competition Policy Com (96) 721 Final [1997] 4 CMLR 519, para 54, 65, 82, 85, Communication from the Commission on the Application of the Community Competition Rules to Vertical Restraints (Follow-up to the Green Paper on Vertical Restraints) Com(98)544 Final [1998] OJ C 365/3, 4, 9, 12, 21, Guidelines on Vertical Restraints [2000] OJ C 291/1, Commission Regulation (EC) No 2790/1999 [1999] OJ L 336/21, and Peeperkorn (1998: 10). On the status of guidelines and recitals in Community law: Korah and O'Sullivan (2002: para 3.2), Cosma and Whish (2003: 50–53) and Joined Cases C-189/02 P *Dansk Rørindustri A/S v Commission* [2005] 28th June, nyr, para. 198–233, particularly para. 211. Compare with Anthony (1992).

[72] Guidelines on Vertical Restraints [2000] OJ C 291/1, para 6, 100–102, 119(1).

de minimis notice reflects this position by using market share to identify market power, and by establishing a rebuttable presumption that undertakings with a market share below 15 per cent lack sufficient market power to generate allocative inefficiency by vertical collusion.[73]

The shift in favour of efficiency occurs not simply because the conflict is recognized. The formal completion of the internal market diminishes the extent to which integration as an ideological constituent of the conception of competition can be supported.[74] Further, the CFI is seen by some as a Court more specialized in competition law than Community law in general, and so less concerned to follow the integration lodestar.[75] Finally, it may no longer be in the gift of the Community to pursue a market integration conception of Article 81 EC. A system of centralized enforcement enabled the Commission to develop and impose the integration conception of competition.[76] The system of competition enforcement was fully decentralized in May 2004.[77] Though the Commission remains responsible for policy, national courts and competition authorities carry out the day-to-day enforcement of Article 81 EC.[78] There is no guarantee that all authorities now competent to apply Article 81EC will share the Commission's conception of competition.[79] Divergent interpretations of Article 81 can be characterized as promoting a dialogue as to what Article 81 is and ought to be about; the Court of Justice will ultimately resolve these debates.[80]

[73] Guidelines on Vertical Restraints [2000] OJ C 291/1, para 8–11, Commission Notice on Agreements of Minor Importance (de Minimis) [2001] OJ C 368/13, para 7(b), 11(2), Communication from the Commission on the Application of the Community Competition Rules to Vertical Restraints (Follow-up to the Green Paper on Vertical Restraints) Com(98)544 Final [1998] OJ C 365/3, 21, Green Paper on Vertical Restraints in EC Competition Policy Com (96) 721 Final [1997] 4 CMLR 519, para 293–300, Riley (1998: 488), Kellaway (1997: 390), Carlin (1996: 286), and Nourry and Lodge (1997: 192–195).

[74] Gerber (1994: 143–145) and Wesseling (2000: 48–50).

[75] Cruz (2002: 100–101).

[76] Article 4(1) and 9(1) of Regulation 17. First Regulation Implementing Articles 81 and 82 of the Treaty [1962] OJ Special Edition 204/62 and Wesseling (2000: 61, 33–48, 62–64).

[77] Council Regulation (EC) No 1/2003 [2003] OJ L 1/1, Paulweber (2000), and Ehlermann (2000).

[78] Wißmann (2000: 137, 141–142), Bourgeois and Humpe (2002: 46–47), Bovis (2001: 102), and Mavroidis and Neven (2001: 159–164).

[79] Mavroidis and Neven (2001: 162), Albors-Llorens (2002: 330–331), Wesseling (2001: 374), Monopolkommission (2000: 35–37), Bovis (2001: 99–100, 102–103), Wißmann (2000: 144–145), Lever (2002: 322), Lever QC and Peretz (1999: para 4.3), Kingston (2001: 348–349), Rodger and Wylie (1997: 486), Schaub (1999: 154), Reeves and Brentford (2000: 77), Siragusa (2000: 1100–1102), Holmes (2000: 69). Compare with Van den Bergh and Camesasca (2001: 133, 149–154).

[80] Venit (2003: 546, 559–566), Fox (2001: 143–145), Wesseling (2001: 367, 373), Wißmann (2000: 147), and Komninos (2001: 226–227). Articles 3, 11, and 16 of Council Regulation (EC) No 1/2003 [2003] OJ L 1/1 seem to exclude the possibility of such debate.

VI. Conclusion

Three constituent values of competitive markets have been specified. First, competition plays a leading role in making the most of society's scarce resources; second, competition is able to contribute to European integration; third, competitive markets epitomize freedom. These values have the potential to conflict and a hierarchical order is established to enable Article 81 EC to be consistently enforced. It is argued that since the purpose of integration is to unlock efficiencies, efficiency cannot be sacrificed to promote integration: efficiency ranks higher than integration.[81] There is then the issue of freedom (competition as a process), efficiency (competition as an outcome), and the hierarchical superiority between the two. Though ordoliberalism values freedom first, the Article 81 EC conception of competition values the ability to make the most of society's scarce resources.[82] The remainder of this book, particularly Chapters 5 and 6, shows how an efficiency conception of competition's value is expressed through Article 81 EC. Chapter 7 returns to consider whether efficiency is the only value pursued.

[81] Wolf (1998).

[82] Case C-250/92 *Gottrup-Klim Grovvareforening and Others v Dansk Landbrugs Grovvareselskab Amba (DLG)* [1994] ECR I-5641, AG Opinion para 13, Amato (1997: 7–10), Black (1997: 146, n 9 plus accompanying text), Bork (1993: 59), Goyder (1998: 572, 586), Korah (1985: 3–7), Korah (2000: 2, 4, 11, 16, 64–65, 364–366), and Wesseling (2001: 359). Compare with Petersmann (1999: 149–150), Hildebrand (2002b: 183–187), and Evans (2001).

3

The meaning of undertaking within Article 81 EC*

I. Introduction

The first indication that Article 81 EC is concerned with efficiency is given by the identity of the addressees of the provision. The nature of the obligation imposed must be an appropriate one to impose on those bound by the provision.[1] Article 81 EC applies to agreements and concerted practices *between undertakings* and to decisions of *associations of undertakings*. Two addressees are specified: (a) undertakings and (b) associations of undertakings.[2] This chapter focuses on the meaning of undertaking. As noted by Advocate General Jacobs, 'the concept of "undertaking" serves a dual purpose in the system of Article [81]. On the one hand—and this function is more obvious—it makes it possible to determine the categories of actors to which the competition rules apply . . . On the other hand, it serves to establish the entity to which a certain behaviour is attributable.'[3] This chapter is concerned with the first purpose.[4] Section II demonstrates that the Court has given a functional definition to the term undertaking: the competition rules apply to entities engaged in economic activity. Section III aims to reveal the meaning attached to economic activity by examining

* This chapter is a slightly revised version of an article published under the same title in the Cambridge Yearbook of European Legal Studies volume 7 (2005).

[1] Fuller (1969: 70–79).

[2] Article 82 EC prohibits 'abuse by one or more *undertakings*'.

[3] Case C-67/96 *Albany International BV v Stichting Bedrijfspensioenfonds Textielindustrie* [1999] ECR I-5751, AG Opinion para 206.

[4] The second purpose is considered by Wils (2000) and receives recent treatment in Joined Cases C-189/02 P etc *Dansk Rørindustri A/S v Commission* [2005] ECR 28th June, nyr, para 104–115. In Levy (2003: para 5.01, n 3, emphasis added), it is stated that 'the concept of undertaking for merger purposes *is similar* to that defined by the Court of Justice under Art 81'. It is likely that it is in the attributional sense that the terms differ.

the jurisprudence of the Court. Section IV considers the need for a functional approach, arguing that it is required to draw a distinction between public and private activity, and some implications of the functional approach adopted. Section V is an attempt to draw conclusions from the enquiry undertaken. Particularly, it should be clear that at the root of the difficulty in determining whether a particular body is an undertaking lies a question of who *should* be bound by the general rules of competition in the EC Treaty. This chapter should thus be seen as a contribution to the debate over the legitimate scope of Community competition law.

II. The functional definition of undertaking

It is clear from the wording of the Treaty that undertakings must not infringe the principles enshrined in Article 81 EC (or Article 82 EC), which is addressed *inter alia* to agreements between undertakings.[5] Undertaking is defined in Article 1 of Protocol 22 EEA as 'any entity carrying out activities of a commercial or economic nature'.[6] The term undertaking is nowhere defined in the EC Treaty.[7] It was recognized at the outset that the Community definition must operate independently of national conceptions of undertaking.[8] In *Commission v Italy* Advocate General Mischo considered that whether a body is to be considered an undertaking turns on 'the industrial and commercial nature of the activity'.[9] The Court of Justice in *Höfner* defines undertakings as 'every entity engaged in an economic activity, regardless of the legal status of the entity and the way in which it is financed'.[10] It is

[5] Wils (2000: 99–100), Cruz (2002: 128) and Case C-67/96 *Albany International BV v Stichting Bedrijfspensioenfonds Textielindustrie* [1999] ECR I-5751, AG Opinion para 206.

[6] Undertaking is defined in Article 80 ECSC as 'any undertaking engaged in production in the coal or the steel industry within the territories referred to . . . and also . . . any undertaking or agency regularly engaged in distribution other than sale to domestic consumers or small craft industries'. Undertaking is defined in Article 196 Euratom as 'any undertaking or institution which pursues all or any of the activities in the territories of Member States within the field specified . . . whatever its public or private legal status'.

[7] Honig *et al.* (1963: 8) write that the Treaty 'assumes that its meaning is understood'.

[8] Deringer (1968: 4–5).

[9] Case 118/85 *Commission v Italian Republic* [1987] ECR 2599, 2610. The Court's first definition of the term 'undertaking' in the context of EC competition law came in Case 170/83 *Hydrotherm Gerätebau* [1984] ECR 2999 para 11, though this definition relates to the attribution function of the term.

[10] Case C-41/1990 *Klaus Höfner and Fritz Elser v Macrotron GmbH* [1991] ECR I-1979, para 21. Also Case 170/83 *Hydrotherm Gerätebau* [1984] ECR 2999 para 11. This definition seems to lean

important to understand that, from this definition, Article 81(1) EC is not addressed to *entities* at all; rather it addresses *activities*. The approach is functional rather than institutional: the functional nature of the question is neatly summarized by Advocate General Jacobs, writing that:

the Court's general approach to whether a given entity is an undertaking within the meaning of the Community competition rules can be described as functional, in that it focuses on the type of activity performed rather than on the characteristics of the actors which perform it... Provided that an activity is of an economic character, those engaged in it will be subject to Community competition law.[11]

An early view expressed by Deringer was that a functional approach 'placed too much emphasis upon the activity, rather than the subject of the activity. [Undertaking] within the meaning of Article [81(1)] cannot be the activity, the doing as such, but only a subject in relation to a certain doing.'[12] However critical, Deringer accepted that a functional approach is taken, so that '[t]he type of activity determines when the legal entity, as an [undertaking], comes within Article [81(1)]'.[13] However, the full implications of the functional approach have yet to be revealed or fully considered.[14] *Entities* are only ever addressed in relation to *activities*. Consequently, sometimes a body is an undertaking and sometimes it is not: there are no bodies that cannot be considered undertakings, only activities that are not considered economic.[15] The functional approach means the relevant question is not *who* is an undertaking but *what* is economic activity?

heavily on one suggested by the Bundekartellamt, reported in Deringer (1968: 5). A similar approach is taken in Swiss competition law, reported in Meinhardt and Waser (2005: 354).

[11] Joined Cases C-264/01, C-306/01, C-354/01 and C-355/01 *AOK Bundesverband v Ichthyol-Gesellschaft Cordes* [2003] ECR I-2493, AG Opinion para 25 (citations omitted), also para 45, and judgment para 58.

[12] Deringer (1968: 5). Compare with the US approach outlined in Haffner (2005).

[13] Deringer (1968: 6). For acceptance of the functional approach: Case 170/83 *Hydrotherm Gerätebau* [1984] ECR 2999 para 11, Case T-319/99 *Federación Nacional de Empresas de Instrumentación Científica, Médica, Técnica y Dental (FENIN) v Commission* [2003] ECR II-357, para 14–19, Case C-343/95 *Diego Cali & Figli Srl v Servizi Ecologici Porto di Genova SpA (SEPG)* [1997] ECR I-1547, para 16–17, AG Opinion para 40, Case C-67/96 *Albany International BV v Stichting Bedrijfspensioenfonds Textielindustrie* [1999] ECR I-5751, AG Opinion para 207, Case C-475/99 *Ambulanz Glöckner v Landkreis Südwestpfalz* [2001] ECR I-8089, AG Opinion para 71–81, Case C-218/00 *Cisal di Battistello Venanzio & Co. v Istituto Nazionale per L'assicurazione Contro Gli Infortuni Sul Lavoro (INAIL)* [2002] ECR I-691, AG Opinion para 48–49, Buendia Sierra (1999: para 1.149–1.152), Goyder (2003: 60–61), Gyselen (2000: 439), and Winterstein (1999: 325).

[14] See Section IV below. Some implications of the functional approach are considered in Hatzopoulos (2002: 710) and Spaventa (2004a: 287).

[15] Drijber (2005: 528).

III. Economic activity

The functional approach established in *Höfner* uses the idea of economic activity to identify the addressees of Community competition law. The task of this section is to determine the constituent elements of economic activity. The Court has considered the question on numerous occasions.[16] For some 'the analysis will always boil down to whether a particular activity could, at least in principle, be carried on by a private undertaking'.[17] In this chapter, it is argued that, though the body of cases awaits authoritative synthesis and articulation by the Court of Justice, it is possible to identify three positive requirements of economic activity. These are that the entity must: offer goods or services to the market; bear the economic or financial risk of the enterprise; and have the potential to make profit from the activity. These cumulative elements are considered in turn.

A. *Offer goods or services in the market place*

From *Commission v Italy* onwards it is clear that economic activity consists of *offering* goods or services on the market.[18] The case arose when the Commission objected to Italy's failure to comply with a directive addressed to undertakings. The Amministrazione Autonoma dei Monopoli di Stato (AAMS) carried out the manufacture and sale of tobacco. However, under Italian law, AAMS had no status separate from that of the state. Italy argued that AAMS did not have to comply with the directive: rather than an undertaking to which the directive was addressed, AAMS was part of the state (a public authority). Ignoring the *institutional* form, the Court considered that AAMS was an undertaking since the *activity*, offering goods or services to the market, was economic.[19]

[16] See Buendia Sierra (1999: para 1.148–1.213) and Case 1006/2/1/01 *Bettercare Group Limited v the Director General of Fair Trading* [2002] CAT 7, para 71–103.

[17] Drijber (2005: 528).

[18] Case 118/85 *Commission v Italian Republic* [1987] ECR 2599, para 7, Case C-343/95 *Diego Cali & Figli Srl v Servizi Ecologici Porto di Genova SpA (SEPG)* [1997] ECR I-1547, para 16, Joined Cases C-180/98 to C-184/98 *Pavel Pavlov v Stichting Pensioenfonds Medische Specialisten* [2000] ECR I-6451, para 75, Case C-475/99 *Ambulanz Glöckner v Landkreis Südwestpfalz* [2001] ECR I-8089, para 19, Case C-218/00 *Cisal di Battistello Venanzio & Co. v Istituto Nazionale per L'assicurazione Contro Gli Infortuni Sul Lavoro (INAIL)* [2002] ECR I-691, para 23, AG Opinion para 38, Case T-319/99 *Federación Nacional de Empresas de Instrumentación Científica, Médica, Técnica y Dental (FENIN) v Commission* [2003] ECR II-357, para 36, Case C-222/04 *Ministero dell'Economia e delle Finanze Cassa v di Risparmio di Firenze SpA* 2005 27th October nyr, AG Opinion para 74 and 77–78, and Montana and Jellis (2003: 112).

[19] Case 118/85 *Commission v Italian Republic* [1987] ECR 2599, para 3, 7.

What constitutes a good has a well-developed meaning under Article 28 EC.[20] Similarly, what constitutes a service has a well-developed meaning under Article 50 EC.[21] These definitions of goods and services are used when deciding whether an entity is offering goods or services to the market.[22] In addition, the Court seems to recognize three activities that do *not* involve the offer of goods or services to the market place. These are work, consumption, and regulation.

1. Work

In *Albany*, the Court and Advocate General considered whether employees were undertakings in relation to their employers.[23] The relevant question is 'how to classify the fact that employees offer *labour* against remuneration.'[24] The Advocate General accepted that '[o]ne could argue that [this] is an economic activity similar to the sale of *goods* or the provision of *services*'.[25] However, Article 28 EC and 50 EC define goods and services; employees provide neither goods nor services to the employer but are engaged in *work* as defined by Article 39 EC. Thus, the employee is not an undertaking in relation to the employer because neither goods nor services are offered.[26] Since goods, services, and work are distinct concepts, to find otherwise would 'necessitate the use of uneasy analogies between the markets for goods and services and labour markets'.[27]

[20] Oliver and Jarvis (2003: para 2.02–2.11). [21] Barnard (2004: 335).

[22] Case T-313/02 *David Meca-Medina and Igor Majcen v Commission* [2004] ECR 30 September nyr, para 42, Case C-41/1990 *Klaus Höfner and Fritz Elser v Macrotron GmbH* [1991] ECR I-1979, AG Opinion at para 19–20, 40. In Case C-205/03 P *FENIN* [2005] 10th November, nyr, AG Maduro considers that the free movement definitions are not always appropriate under the competition rules (para 48–51).

[23] As Jones and Sufrin (2004: 120) point out, the employee/employer relationship is only indirectly considered. At any rate, the case is read as rejecting the idea that Community competition law applies to the employee/employer relationship: Case C-67/96 *Albany International BV v Stichting Bedrijfspensioenfonds Textielindustrie* [1999] ECR I-5751, para 46–60, AG Opinion at para 131–194, 209–217 and Boni and Manzini (2001).

[24] Case C-67/96 *Albany International BV v Stichting Bedrijfspensioenfonds Textielindustrie* [1999] ECR I-5751, AG Opinion para 211, emphasis added.

[25] Case C-67/96 *Albany International BV v Stichting Bedrijfspensioenfonds Textielindustrie* [1999] ECR I-5751, AG Opinion para 212, emphasis added.

[26] On the competition law status of the employee/employer relationship in various jurisdictions see Case C-67/96 *Albany International BV v Stichting Bedrijfspensioenfonds Textielindustrie* [1999] ECR I-5751, AG Opinion at para 80–111 and Brunn and Hellsten (2001).

[27] Case C-67/96 *Albany International BV v Stichting Bedrijfspensioenfonds Textielindustrie* [1999] ECR I-5751, AG Opinion para 216.

2. Consumption

Whilst the offeror of goods or services is an undertaking, the recipient of goods or services is only an undertaking (engaged in economic activity) if the goods or services purchased form an essential element of goods or services subsequently offered to the market place. This position finds support in *AOK*: the Advocate General reported 'The Oberlandesgericht suggest[ion] that purchasing may amount to an economic activity whether or not the entity which purchases is itself active on another market for which the goods or services purchased constitute an input. The Commission, the appellants and the German Government all contend otherwise.'[28] The immunity of consumption was accepted at the outset, so that Deringer writes that the competition rules do not apply to activity 'which has the sole objective of meeting personal needs'.[29] When the purchase is not an input for further downstream activity the purchaser is not engaged in economic activity.[30] Article 81 EC cannot then apply, as there is a need for action *between* undertaking*s*.

In *FENIN*, it was argued that the Spanish Health Service was involved in economic activity when it *purchased* goods and services, even when it did not offer goods and services downstream.[31] The Court of First Instance rejected the contention, finding that:

an organisation which purchases goods—even in great quantity—*not for the purpose of offering goods and services as part of an economic activity*, but in order to use them in the context of a different activity, such as one of a purely social nature, does not act as an undertaking simply because it is a purchaser in a given market . . . if the activity for which that entity purchases goods is not an economic activity, it is not acting as an undertaking for the purposes of Community competition law and is therefore not subject to the prohibitions laid down in Articles 81(1) EC and 82 EC.[32]

[28] Joined Cases C-264/01, C-306/01, C-354/01 and C-355/01 *AOK Bundesverband v Ichthyol-Gesellschaft Cordes* [2003] ECR I-2493, AG Opinion para 46.

[29] Deringer (1968: 8).

[30] Guidelines on Vertical Restraints [2000] OJ C 291/1, para 24. Compare with Case 1006/2/1/01 *Bettercare Group Limited v the Director General of Fair Trading* [2002] CAT 7, para 46, 53, 264, Note (2004), Lasok (2004: 383), and Drijber (2005).

[31] Case T-319/99 *Federación Nacional de Empresas de Instrumentación Científica, Médica, Técnica y Dental (FENIN) v Commission* [2003] ECR II-357.

[32] Case T-319/99 *Federación Nacional de Empresas de Instrumentación Científica, Médica, Técnica y Dental (FENIN) v Commission* [2003] ECR II-357, para 37, emphasis added. The Court does not restrict the excluded use to consumption, and so it may be more accurate to use the term dissipation. This is confirmed by AG Maduro in Case 205/03 P *FENIN* para 63, 64 and 66. At the time of writing the ECJ has yet to rule on the issue. See note 40 below.

It is clear that consumption is not economic activity because the element of offering goods or services to the market is lacking. However, purchasing activities must still be classified as either related or unrelated to the offer of goods or services downstream.[33] A difficulty in making this determination is that there are varying degrees to which the purchased good or service is a necessary input to a good or service subsequently offered. In *Pavlov*, the Court considered the argument that medical practitioners were purchasing supplementary pension benefits not as undertakings but as consumers.[34] The Court rejected the argument because the purchase of the pension was 'closely linked to the practice of his profession'.[35] How to determine whether the link between purchased goods or services is sufficiently proximate to goods or services subsequently resold is a difficult question of fact.

More difficulty in determining whether the purchase is for consumption or resale arises when the good or service is purchased by a body entrusted with a social service obligation, which then provides the good or service to an end consumer free of charge or at a discount rate.[36] Case law under Article 50 EC would suggest that the activity remains economic, though there are inconsistencies.[37] This fits with the oft-repeated mantra under Article 81 EC that activity is economic independently of 'the way in which it is financed'.[38] If the body providing remuneration is considered a purchaser for resale, then in relation to the ultimate recipient the purchaser offers goods or services, even if at 100 per cent discount. This is how the UK Competition Appeal Tribunal (CAT) characterized the activity of a social service provider under a statutory duty to both provide social services (for which it received an annual governmental grant) and to recover the full cost of providing the services from the recipient, subject to ability to pay.[39]

[33] Case T-319/99 *FENIN v Commission* [2003] ECR II-357, para 36 and Office of Fair Trading (2004a). In Case C-222/04 *Ministero dell' Economia e delle Finanze v Cassa di Risparmio di Firenze SpA*, AG Jacobs considers the need to offer goods and services too tightly constrains the competition provisions and proposes a looser test (para 78). This contrasts with the approach taken by AG Maduro in Case C-205/03 P *FENIN* (para 26–27, 58–69), see note 32 above.

[34] Joined Cases C-180/98 to C-184/98 *Pavel Pavlov v Stichting Pensioenfonds Medische Specialisten* [2000] ECR I-6451, para 78.

[35] Ibid, para 79.

[36] Buendia Sierra (1999: para 1.156–1.157).

[37] Case 352/85 *Bond v the Netherlands* [1988] ECR 2085, para 16, Case C-109/92 *v Landeshauptstadt Hannover* [1993] ECR I-5473, para 17, Case 263/86 *Belgium v Humbel* [1988] ECR 5365, para 17–19.

[38] Case C-41/1990 *Klaus Höfner and Fritz Elser v Macrotron GmbH* [1991] ECR I-1979, para 20. Compare with Oliver (2000: 481–482).

[39] Case 1006/2/1/01 *Bettercare Group Limited v the Director General of Fair Trading* [2002] CAT 7, para 46, 49, 110–115, 136–137, 140–150, 155–156, 183–188, 201, 264. Though the case is

Conversely, in *FENIN*, the Court of First Instance considered that there was no resale, so that the purchase was as a consumer rather than as an undertaking.[40] The complexity of characterizing the nature of any transfer of benefits by the remunerator to a third party should be clear.

3. *Regulation*

A third category of activity that does not entail offering goods or services to the market place is market regulation.[41] In *Bodson*, French law required regional authorities to regulate the provision of various aspects of funeral services.[42] Some authorities achieved their regulatory objectives by granting exclusive concessions to carry out funeral services in their particular region. Advocate General Vilaça and the Court considered that the regional authority was not an undertaking as they were not engaged in economic activity but were instead carrying out a regulatory function.[43] Since regulation does not involve offering goods or services to the market, it is not economic activity, even though its *raison d'être* is to affect how markets operate.[44]

There are those that question the extent to which regulating economic activity can be separated from economic activity itself.[45] There are those that argue that, because regulation can have anti-competitive effects,

decided under the UK Competition Act 1998, the meaning of 'undertaking' is intended to be the same as in EC competition law: Case 1006/2/1/01 *Bettercare Group Limited v the Director General of Fair Trading* [2002] CAT 7, para 29–34. The controversy caused is noted by Bright and Currie (2003), Skilbeck (2002), Prosser (2005: 54–57), and Office of Fair Trading (2004a). The OFT subsequently adopted a non-infringement decision in Case CE/1836–02 *Bettercare Group Ltd/ North & West Belfast Health & Social Services Trust (Remitted Case)* [2003] 18 December, para 22, 24–39 because the price was set by the state and in setting the price the state was not engaged in economic activity. The case can be seen as turning on the meaning of undertaking in its attributional sense.

[40] See text accompanying note 32 above and Prosser (2005: 129–130). The Court of First Instance did not address the fact that sometimes there was clearly resale, Case T-319/99 *FENIN v Commission* [2003] ECR II-357, para 19. In Case 205/03 P *FENIN* (para 54–57) AG Maduro asked that the case be refered back to the CFI to clarify whether there was resale as a factual matter.

[41] Case 5/79 *Procureur General v Hans Buys* [1979] ECR 3203, para 30, Cases 1002/2/1/01(IR), 1003/2/1/01, 1004/2/1/01 *GISC* [2001] CAT 4, para 142, 156, Buendia Sierra (1999: 55–56), Gagliardi (2000: 360, 365), and Prosser (2005: 128).

[42] The regulated aspects were the carriage of the body after it has been placed in the coffin, the provision of hearses, coffins and external hangings of the house of the deceased, conveyances for mourners, the equipment and staff needed for burial and exhumation and cremation.

[43] Case 30/87 *Bodson v Pompes Funèbres des Régions Libérées SA* [1988] ECR 2479, para 18, AG opinion para 94.

[44] Marenco (1987: 421–423).

[45] Shaw (1988: 423) and Cases 1002/2/1/01(IR), 1003/2/1/01, 1004/2/1/01 *GISC* [2001] CAT 4, para 245–247.

regulatory standards should be subjected to Community competition law.[46] This approach is supported by the Court writing that Article 81(1) EC 'applies ... in so far as ... activities ... are calculated to produce the results which it aims to suppress'.[47] Moreover, the claim that regulation is not economic activity seems to be contradicted by *Wouters*, a case in which the Court considers that when an association of undertakings 'acts as the *regulatory body* of a profession' that practice 'constitutes an *economic activity*'.[48]

It should first be noted that in *Wouters* the Court is considering the activity of an *association* of undertakings; associations of undertakings *may* fall within the scope of the competition rules even though their activity is *not* economic.[49] The Court treated the association as an association rather than as an undertaking, so economic activity was unnecessary. Second, whilst regulatory regimes must be efficient in attaining their goals, the goals of a regulatory regime need not be efficiency. As Prosser notes, '[m]arkets are never free, being constructed through, and dependent on, different kinds of legal (and social) structures'.[50] A regulatory regime that does not pursue efficiency cannot be seen as anti-competitive since efficiency is not necessarily the criterion by which regulatory regimes should be assessed: the competition rules are simply an inappropriate lens through which to view regulation.[51] Finally, the argument that regulation does not involve the offer of goods or services to the market; that as such it *is not* economic activity; that those engaged in it do not act as undertakings and are *not* subject to Community competition law, is made more forceful once it is realized that regulatory activity is subject to *appropriate* Community law.[52] In *Walrave and Koch*, the free-movement rules are seen as applicable to all entities that promulgate '*rules of any other nature aimed at regulating* in a collective manner gainful employment and the provision of services'.[53]

[46] Gagliardi (2000) and Hervey (2000: 39–40).

[47] Joined cases 209 to 215 and 218/78 *Heintz Van Landewyck SARL and Others v Commission (FEDETAB)* [1980] ECR 3125, para 88, also Case C-244/94 *Fédération Française des Sociétés d'assurance v Ministère de l'agriculture et de la Pêche* [1995] ECR I-4013, para 21, Case 1006/2/1/01 *Bettercare Group Limited v the Director General of Fair Trading* [2002] CAT 7, para 244–249, Montana and Jellis (2003: 114) and Skilbeck (2002: 261–262).

[48] Case C-309/99 *Wouters v Algemene Raad Van de Nederlandse Orde Van Advocaten (Raad Van de Balies Van de Europese Gemeenschap, Intervening)* [2002] ECR I-1577, para 58, emphasis added. Also Case T-193/02 *Laurent Piau v Commission* 26 January 2005, nyr, para 77–78.

[49] See Section IV.B below.

[50] Prosser (2005: 1).

[51] Compare with Whish (2003: 120–123) and Nihoul (2004: 103–104). On criteria for assessing regulatory regimes see Yeung (2004: 29–55).

[52] Drijber (2005: 530–531).

[53] Case 36–74 *Walrave & Koch v Association Union Cycliste Internationale* [1974] ECR 1405, para 17, emphasis added, confirmed in Case 13–76 *Gaetano Dona v Mario Mantero* [1976] ECR 1333,

Regulation, whether by state or non-state actors, is covered by what can be seen as EC public law because of the 'quasi-government status' of 'the ultimate regulatory body', which can thus be seen to perform 'State-like functions'.[54] Rather than involving economic activity, regulation involves the exercise of official authority (public power or *imperium*). Advocate General Tesauro has reported that:

> the Court has preferred not to define that concept [official powers] in abstract terms, but has instead followed 'the path marked out by Advocate General Mayras in his Opinion in the *Reyners* case, according to whom "official authority is that which arises from the sovereignty and majesty of the State for him who exercises it, it implies the power of enjoying the prerogatives outside the general law, privileges of official power and powers of coercion over citizens" '.[55]

The Court has consistently held that 'bodies that exercise an activity typical of a public authority ... do not constitute undertakings and are not therefore subject to the Community rules on competition'.[56] By applying public law to the regulatory *activity* of an entity that is *institutionally* private, 'public and private regulation are put on the same footing by the Court of Justice'.[57] It seems clear that if the regulatory rules at issue in *Wouters* were adopted by the state itself they would be subject to the EC free-movement

para 17. See also Case T-313/02 *David Meca-Medina and Igor Majcen v Commission* [2004] ECR 30 September nyr, para 44–47, on appeal as Case C-519/04 P, Weatherill (2005) and Gregory (2005). Weatherill (1989: 63–66) considers *Walrave* as authority for the proposition that the free-movement rules can be applied to operations in the private sphere, as opposed to the view taken here that it established that regulation is inherently public sphere activity. van den Bogaert (2002: 126–128) notes that Case 251/83 *Haug-Adrion* [1984] ECR 4277 supports the Weatherill view, but that it is not followed in Case C-415/93 *ASBL v Jean-Marc Bosman, Royal Club Liégeois SA v Jean-Marc Bosman and Others and UEFA v Jean-Marc Bosman* [1995] ECR I-4921.

[54] van den Bogaert (2002: 126). On the free-movement rules as EC public law see Section IV.A below.

[55] Case C-364/92 *SAT Fluggesellschaft mbH v Eurocontrol* [1994] ECR I-43, AG Opinion para 9, citing Case 2/74 *Reyners v Belgian State* [1974] ECR 631, Case 149/79 *Commission v Belgium* [1980] ECR 3881 Case 149/79 *Commission v Belgium* [1982] ECR 1845 Case 307/84 *Commission v France* [1986] ECR 1725, Joined Cases 231/87 and 129/88 *Ufficio distrettuale delle imposte dirette di Fiorenzuola d'Arda* [1989] ECR 3233. This definition of official authority is used by AG Jacobs in Case C-41/1990 *Klaus Höfner and Fritz Elser v Macrotron GmbH* [1991] ECR I-1979, para 22. Compare with the use of 'authority' and 'imperium' in Oliver (2000: 481) and Allison (1997: 82) respectively.

[56] Case C-343/95 *Diego Cali & Figli Srl v Servizi Ecologici Porto di Genova SpA (SEPG)* [1997] ECR I-1547, AG Opinion, para 41. Also Case 30/87 *Bodson v Pompes Funèbres des Régions Libérées SA* [1988] ECR 2479, para 18, Winterstein (1999: 325–327), Gyselen (2000: 439–440), Louri (2002: 160–164), and Montana and Jellis (2003: 111).

[57] van den Bogaert (2002: 125). Milner-Moore (at text accompanying note 25–26) writes that dealing with the conduct of private actors under the free movement rules only when they are

rather than the EC competition rules; this shows some recognition that the regulatory function is not economic.[58]

Since regulation does not involve the offer of goods or services to the market, it is important to determine whether activity is regulatory.[59] There is no clear test and particular problems are caused by the activities of self-regulatory bodies. The Court initially took a functional approach to determine whether activity is regulatory.[60] So, for example, UEFA is said to be engaged in regulatory activity for the purpose of football transfer rules and nationality restrictions, but its other activities are seen as economic.[61] However, an exclusively functional approach seems to be rejected.[62] Whilst the question of whether activity is regulatory remains functional, there is increased reliance on institutional factors to assist in the determination. Whether an entity is composed of those chosen for their independent expertise; the state's involvement in the entity; whether the entity is subject to substantive obligations to operate in the public interest; and whether there is consultation before action, have all been considered.[63]

engaged in regulatory activity is 'superficially attractive'. However, compare with Joined cases 266 and 267/1987 *The Queen v Royal Pharmaceutical Society of Great Britain, ex Parte Association of Pharmaceutical Importers* [1989] ECR 1295, para 14–15 and Hatzopoulos (2002: 709).

[58] Case C-55/94 *Reinhard Gebhard v Consiglio dell'Ordine degli Avvocati e Procuratori di Milano* [1996] ECR I-4165 and Lonbay (1996).

[59] Joined Cases C-264/01, C-306/01, C-354/01 and C-355/01 *AOK Bundesverband v Ichthyol-Gesellschaft Cordes* [2003] ECR I-2493, AG Opinion para 52.

[60] Case C-415/93 *ASBL v Jean-Marc Bosman, Royal Club Liégeois SA v Jean-Marc Bosman and Others and UEFA v Jean-Marc Bosman* [1995] ECR I-4921, para 83–84, van den Bogaert (2002: 124–126), Milner-Moore (at text accompanying notes 111–125). Compare with Whish (2003: 98).

[61] Case C-415/93 *ASBL v Jean-Marc Bosman, Royal Club Liégeois SA v Jean-Marc Bosman and Others and UEFA v Jean-Marc Bosman* [1995] ECR I-4921, para 82, 87, Joint Selling of the Commercial Rights of the UEFA Champions League [2003] OJ L 291/25, para 106, Case CP/0871/01 *Price-Fixing of Replica Football Kit* [2003] 1 August, para 305, and references cited at note 158 below.

[62] Case C-309/99 *Wouters v Algemene Raad Van de Nederlandse Orde Van Advocaten (Raad Van de Balies Van de Europese Gemeenschap, Intervening)* [2002] ECR I-1577, AG Opinion para 76, 85–86.

[63] Case 249/1981 *Commission v Ireland* [1982] ECR 4005, para 15, Case 222/1982 *Apple and Pear Development Council v K J Lewis Ltd* [1983] ECR 4083, para 2–7, 17, Joined Cases C-180/98 to C-184/98 *Pavel Pavlov v Stichting Pensioenfonds Medische Specialisten* [2000] ECR I-6451, para 87–88, Case C-309/99 *Wouters v Algemene Raad Van de Nederlandse Orde Van Advocaten (Raad Van de Balies Van de Europese Gemeenschap, Intervening)* [2002] ECR I-1577, para 51, 58–71, AG Opinion at para 70, 74–75, Joined Cases C-264/01, C-306/01, C-354/01 and C-355/01 *AOK Bundesverband v Ichthyol-Gesellschaft Cordes* [2003] ECR I-2493, AG Opinion para 53, Cruz (2002: 148–149, 154), Schepel (2002: 36–38, 44–47), Snell (2002: 218–221), Stiglitz (2000: 13–14), and Vossestein (2002: 848–849).

B. *Bear economic (financial) risk*

In addition to offering goods or services to the market, the entity must bear the financial risk of the enterprise going awry. In *Pavlov*, medical specialists were considered undertakings because they 'assume the financial risks attached to the pursuit of their activity'.[64] In *Wouters*, after considering the service offered to the market by lawyers, the Court felt it important to the characterization as economic activity that 'they bear the financial risks attaching to the performance of those activities since, if there should be an imbalance between expenditure and receipts, they must bear the deficit themselves'.[65] Risk bearing is an essential component of the concept of economic activity used to determine the addressee of the competition rules: the absence of risk bearing prevents activity being seen as economic. In *Poucet and Pistre* individual entities providing social insurance *did not* bear the risk of failure: risk was spread across the sector so that 'those in surplus contribute to the financing of those with structural financial difficulties'.[66] The absence of risk bearing is one factor that prevented the entities being seen as engaged in economic activity in *AOK*.[67]

The absence of risk bearing has been used to explain why employees are not engaged in economic activity.[68] In *Albany* Advocate General Jacobs considered that '[d]ependent labour is by its very nature the opposite of the independent exercise of an economic or commercial activity. *Employees normally do not bear the direct commercial risk of a given transaction.*'[69] Whilst the employee does not bear risk in relation to the outside world, they seem to bear risk in relation to the employer. In this relationship, it would seem that 'employees are acting autonomously and in their own

[64] Joined Cases C-180/98 to C-184/98 *Pavel Pavlov v Stichting Pensioenfonds Medische Specialisten* [2000] ECR I-6451, para 73.

[65] Case C-309/99 *Wouters v Algemene Raad Van de Nederlandse Orde Van Advocaten (Raad Van de Balies Van de Europese Gemeenschap, Intervening)* [2002] ECR I-1577, para 48, AG Opinion para 51.

[66] Joined Cases C-159/91 and C-160/91 *Christian Poucet v Assurances Generales de France (AGF) and Caisse Mutuelle Regionale du Languedoc-Roussillon (Camulrac) and Daniel Pistre v Caisse Autonome Nationale de Compensation de l'Assurance Vieillesse des Artisans (Cancava)* [1993] ECR I-637, para 12, cited and accepted by the UK CAT in Case 1006/2/1/01 *Bettercare Group Limited v the Director General of Fair Trading* [2002] CAT 7, para 239–240. Contra: Case C-244/94 *Fédération Française des Sociétés d'Assurance v Ministère de l'Agriculture et de la Pêche* [1995] ECR I-4013, AG Opinion para 18.

[67] Joined Cases C-264/01, C-306/01, C-354/01 and C-355/01 *AOK Bundesverband v Ichthyol-Gesellschaft Cordes* [2003] ECR I-2493, para 53.

[68] See also Section III.A.1 above.

[69] Case C-67/96 *Albany International BV v Stichting Bedrijfspensioenfonds Textielindustrie* [1999] ECR I-5751, AG Opinion para 215, emphasis added. This reasoning is criticized in Nihoul (2002).

right'.[70] It may well be that the risk that the employer will cease to demand the employee's goods or services does not bring the relationship under competition law scrutiny. However, it is clear that undertaking is being used in its attributional sense and that risk is used to confer *responsibility*: employees are non-undertakings because all of their conduct is attributable to their employers.[71] As explained by Advocate General Colomer in *Becu*:

[i]t is that ability to take on financial *risks* which gives an operator sufficient significance to be capable of being regarded as an entity genuinely engaged in trade, that is to say to be regarded as an undertaking. In other words, recognition as an 'undertaking' requires, at least, the existence of an identifiable centre to which economically significant decisions can be *attributed*. For that reason, employees do not constitute undertakings.[72]

C. *Potential to make profit*

The third element of economic activity is that when bearing the financial risk of offering goods or services to the market place there is at least the potential to make a profit. The juridical basis of this element lies in the Court finding activity economic that 'has not always been, and is not necessarily, carried out by public entities'.[73] In *AOK* Advocate General Jacobs reported that:

[i]n assessing whether an activity is economic in character, the basic test appears ... to be whether it *could*, at least in principle, be carried on by a private undertaking in order to make profits. If there were no possibility of a private undertaking carrying

[70] Case C-67/96 *Albany International BV v Stichting Bedrijfspensioenfonds Textielindustrie* [1999] ECR I-5751, AG Opinion para 210.

[71] Joined Cases 40/73 to 48/73, 50/73, 54/73 to 56/73, 111/73, 113/73, and 114/73 *Suiker Unie and Others v Commission* [1975] ECR 1663, para 539. The definition of employee (worker) under Article 39 EC given in Case 66/85 *Lawrie-Blum v Land Baden-Wurttemberg* [1986] ECR 2121, para 17 requires that a person provides services 'for and under the direction of another'. Risk for the purpose of attribution (thus responsibility) is central to determining whether agents are covered by Community competition rules: Whish (2003: 586–588).

[72] Case C-22/98 *Jean Claude Becu* [2001] ECR I-5665, AG Opinion para 53–54, emphasis added.

[73] Case C-41/1990 *Klaus Höfner and Fritz Elser v Macrotron GmbH* [1991] ECR I-1979, para 22. The potential to make profit is explicitly used by numerous Advocates General, in particular in Joined Cases C-159/91 and C-160/91 *Christian Poucet v Assurances Generales de France (AGF) and Caisse Mutuelle Regionale Du Languedoc–Roussillon (CAMULRAC) and Daniel Pistre v Caisse Autonome Nationale de Compensation de l'Assurance Vieillesse des Artisans (CANCAVA)* [1993] ECR I-637, AG Opinion para 7–8, Case C-364/92 *SAT Fluggesellschaft mbH v Eurocontrol* [1994] ECR I-43, AG Opinion para 9, Case C-244/94 *Fédération Française des Sociétés d'Assurance v Ministère de l'Agriculture et de la Pêche* [1995] ECR I-4013, AG Opinion para 11, Case C-343/95 *Diego Cali & Figli Srl v Servizi Ecologici Porto di Genova SpA (SEPG)* [1997] ECR I-1547, AG Opinion para 32, Case C-67/96 *Albany International BV v Stichting Bedrijfspensioenfonds Textielindustrie* [1999] ECR I-5751, AG Opinion para 311, Case T-128/98 *Aéroports de Paris* [2000]

on a given activity, there would be no purpose in applying the competition rules to it.[74]

It is not necessary actually to make profit, nor is it necessary to have a profit-making motive.[75] All that is required is that the potential exists to make profit from the activity. The surest way to determine whether activities can be carried out to make profit is to look to the market, in either the particular territory in question or other Member States or third states. This is clear from *Höfner* itself. Here German law regulated the provision of recruitment consultancy services by only allowing them to be provided by a designated state body (the Bundesanstalt für Arbeit). Germany argued that the Bundesanstalt für Arbeit was not engaged in economic activity because it provided the recruitment services free of charge.[76] However, the state recruitment consultancy was unable to meet demand for its services. 72 per cent of vacancies were instead filled by a grey market of between 700 and 800 private recruitment consultants in an industry worth between €383 million and €614 million.[77] This makes it clear that even though the Bundesanstalt für Arbeit did not profit from the goods and services offered to the market, there was the *potential* to make profit from the activity. The Advocate General and the Court thus saw recruitment consultancy as economic activity and the Bundesanstalt für Arbeit as an undertaking.[78]

If it is clear that activity is not economic when it is impossible to profit from it, it can further be argued that the Court has accepted two activities

ECR II-3929, para 124, Case C-218/00 *Cisal (INAIL)* [2002] ECR I-691, AG Opinion para 38, Louri (2002: 145–147) and Winterstein (1999: 325). In Case C-222/04 *Ministero dell' Economia e delle Finanze v Cassa di Risparmio di Firenze SpA* AG Opinion para 78, AG Jacobs considers that this requirement *ought* to be determinative of whether or not activity is economic.

[74] Joined Cases C-264/01, C-306/01, C-354/01 and C-355/01 *AOK Bundesverband* [2003] ECR I-2493, AG Opinion para 27, emphasis added, citations omitted.

[75] Joined Cases 209 to 215 and 218/78 *Van Landewyck v Commission* [1980] ECR 3125, para 88, Case C-244/94 *Fédération Française des Sociétés d'Assurance* [1995] ECR I-4013, para 21 and Montana and Jellis (2003: 112). However, compare with Case C-67/96 *Albany International BV* [1999] ECR I-5751, para 74, AG Opinion para 214, 311, 338, and Joined Cases C-264/01, C-306/01, C-354/01 and C-355/01 *AOK Bundesverband* [2003] ECR I-2493, para 47, 51. On irrelevance of profit in the Article 50 EC conception of economic activity: Davies (2002: 29–30).

[76] Case C-41/1990 *Klaus Höfner and Fritz Elser v Macrotron GmbH* [1991] ECR I-1979, para 19, AG Opinion para 1–10.

[77] Case C-41/1990 *Klaus Höfner and Fritz Elser v Macrotron GmbH* [1991] ECR I-1979, AG Opinion para 14–17.

[78] Case C-41/1990 *Klaus Höfner and Fritz Elser v Macrotron GmbH* [1991] ECR I-1979, para 21, AG Opinion para 40. Also Case C-67/96 *Albany International BV* [1999] ECR I-5751, para 83–84 and Case C-475/99 *Ambulanz Glöckner v Landkreis Südwestpfalz* [2001] ECR I-8089, para 14, 20, AG Opinion para 68. This approach to determining whether profit can be made from an activity is given consideration in Case C-205/03 P *FENIN*, AG Opinion para 11–12.

from which profit can never be made. These are redistribution and the provision of public goods: the inherent inability to profit from such activity is explained below.

1. Redistribution

The Court of Justice often speaks about entities pursuing 'an exclusively social objective' that are said not to be engaged in economic activity.[79] What makes an activity exclusively social is 'solidarity'.[80] In *Poucet and Pistre*, the Court found that '[s]olidarity entails the redistribution of income between those who are better off and those who, in view of their resources . . . would be deprived'.[81] As defined by Advocate General Fennelly, '[s]ocial solidarity envisages the *inherently uncommercial* act of *involuntary subsidization* of one social group by another'.[82] *Inter alia*, redistribution can occur between the rich and the poor, the healthy and the sick, the young and the old, low risk and high risk, or active workers and retired workers.[83] How the logic of redistribution differs from the logic of the marketplace is analysed in *Bosman* by Advocate General Lenz, who recognized that sporting competition was unlike market competition, finding that:

football is characterized by the mutual economic dependence of the clubs . . . Each club thus needs the other one in order to be successful. For that reason each club has

[79] Joined Cases C-264/01, C-306/01, C-354/01 and C-355/01 *AOK Bundesverband v Ichthyol-Gesellschaft Cordes* [2003] ECR I-2493, para 47.

[80] It is not clear whether social and *exclusively* social are the same thing. Here the view is taken that the two are distinct and that profit is only impossible when pursuing an *exclusively social* objective. Further, exclusively social is a conclusion drawn when a measure is based on the principle of solidarity. Solidarity is distinguished from non-profit-making and social purpose in Joined Cases C-159/91 and C-160/91 *Christian Poucet v Assurances Generales de France (AGF) and Caisse Mutuelle Regionale du Languedoc-Roussillon (CAMULRAC) and Daniel Pistre v Caisse Autonome Nationale de Compensation de l'Assurance Vieillesse des Artisans (CANCAVA)* [1993] ECR I-637, AG Opinion para 9. On solidarity, see Menendez (2003), Winterstein (1999: 327–331), Louri (2002: 169–172) and Spaventa (2004a: 275, 284–285).

[81] Joined Cases C-159/91 and C-160/91 *Christian Poucet v Assurances Generales de France (AGF) and Caisse Mutuelle Regionale du Languedoc-Roussillon (CAMULRAC) and Daniel Pistre v Caisse Autonome Nationale de Compensation de l'Assurance Vieillesse des Artisans (CANCAVA)* [1993] ECR I-637, para 10, also AG Opinion para 9–11.

[82] Case C-70/95 *Sodemare SA v Regione Lombardia* [1997] ECR I-3395, AG Opinion para 29, emphasis added.

[83] Jones and Sufrin (2004: 534) and Joined Cases C-159/91 and C-160/91 *Christian Poucet v Assurances Generales de France (AGF) and Caisse Mutuelle Regionale du Languedoc-Roussillon (CAMULRAC) and Daniel Pistre v Caisse Autonome Nationale de Compensation de l'Assurance Vieillesse des Artisans (CANCAVA)* [1993] ECR I-637, para 11. Redistribution from an identified affluent to an identified needy group was lacking in Case C-244/94 *Fédération Française des Sociétés d'Assurance* [1995] ECR I-4013, AG opinion at para 19.

an interest in the health of the other clubs. The clubs in a professional league thus do not have the aim of excluding their competitors from the market. Therein lies . . . a significant difference from the competitive relationship between undertakings in other markets.[84]

The feature unifying activities of 'solidarity' is that they are redistributive.[85] It is self-evidently impossible to profit from redistribution, which involves unilateral transfer as opposed to exchange, and which requires the limits of altruism to be overcome by compulsion. Redistribution must not only be the purpose, it must also be the effect.[86] In *Albany*, after finding that a pension scheme was redistributive between those currently employed and those currently retired, Advocate General Jacobs could not:

see any—even theoretical—possibility that without State intervention private undertakings could offer on the markets a pension scheme based on the redistribution principle. Nobody would be prepared to pay for the pensions of others without a guarantee that the next generation would do the same.[87]

The Advocate General 'consequently [had] some difficulty with the view that the activities of such a scheme could be of an economic nature'.[88]

Whilst it is clearly impossible to profit from redistributive activity, the determination of whether activity is redistributive has proved problematic. In *Cisal* the Advocate General considered that the existence of redistribution elements:

may be so fundamental and predominant that as a matter of principle no private [body] can offer that type of [good or service] on the market. On the other hand,

[84] Case C-415/93 *ASBL v Jean-Marc Bosman, Royal Club Liégeois SA v Jean-Marc Bosman and Others and UEFA v Jean-Marc Bosman* [1995] ECR I-4921, AG Opinion para 227.

[85] Redistribution is seen as the hallmark of solidarity in Case C-70/95 *Sodemare SA v Regione Lombardia* [1997] ECR I-3395, para 29, Case C-218/00 *Cisal di Battistello Venanzio & Co. v Istituto Nazionale per L'assicurazione Contro Gli Infortuni Sul Lavoro (INAIL)* [2002] ECR I-691, AG Opinion para 56, 59–60, Joined Cases C-264/01, C-306/01, C-354/01 and C-355/01 *AOK Bundesverband v Ichthyol-Gesellschaft Cordes* [2003] ECR I-2493, AG Opinion para 32, Case 1006/2/1/01 *Bettercare Group Limited v the Director General of Fair Trading* [2002] CAT 7, para 51, 239–240, and Joint Selling of the Commercial Rights of the UEFA Champions League [2003] OJ L 291/25, para 164–167. It is accepted as the hallmark of solidarity by Boni and Manzini (2001: 240–241), Hatzopoulos (2002: 684, 711–712), Van den Bergh and Camesasca (2000: 492, 505) and Prosser (2005: 102–103).

[86] Case C-415/93 *ASBL v Jean-Marc Bosman, Royal Club Liégeois SA v Jean-Marc Bosman and Others and UEFA v Jean-Marc Bosman* [1995] ECR I-4921, AG Opinion para 147, 218–234 and Weatherill (1996: 994, 999, 1004–1006, 1012–1017).

[87] Case C-67/96 *Albany International BV v Stichting Bedrijfspensioenfonds Textielindustrie* [1999] ECR I-5751, AG Opinion para 338.

[88] Case C-67/96 *Albany International BV v Stichting Bedrijfspensioenfonds Textielindustrie* [1999] ECR I-5751, AG Opinion para 338.

they may not go so far as to prevent its activities from being regarded as economic activities[89]

One view is that profit cannot be made from redistribution when the winners and losers are identifiable *ex ante* but it is possible to profit from redistribution when the winners and losers can only be identified *ex post*, eg, insurance.[90] Prosser considers that whether or not competition law applies turns on whether an entity participates in or replaces markets.[91] A more cynical view is that the task of identifying redistributive activity that is non-economic is a fool's errand; redistributive activities are economic, but the Court has treated them as non-economic in order to shield them from the strictures of Community competition law.[92] This view is supported by cases that consider redistribution claims in the assessment of justification rather than jurisdiction.[93] However, a number of elements possessed by redistributive activity can be identified: these are (a) compulsion; (b) control over cost; (c) control over price; and (d) absence of link between cost and price. These features, which are discussed in turn below, are more readily identified in Advocates General opinions than in decisions of the Court. A question mark must thus hang over whether the elements are ever relied upon by the Court and relied upon in the same way. Additionally, it remains unclear whether the elements are complete; cumulative; alternative, or simply a few of a number of factors to which weight is attached in determining whether profit can be made from an activity.[94]

[89] Case C-218/00 *Cisal (INAIL)* [2002] ECR I-691, AG Opinion para 67. Also Joined Cases C-264/01, C-306/01, C-354/01 and C-355/01 *AOK Bundesverband* [2003] ECR I-2493, AG Opinion para 35.

[90] For a discussion of redistributive activities falling within the scope of Community competition law see Hancher and Buendia Sierra (1998) and Abbamonte (1998). Redistribution is also considered in Chapter 6, Section IV.C. [91] Prosser (2005: 125–132). Also Case C-205/03 P *FENIN*, AG Opinion para 12–14.

[92] Buendia Sierra (1999: 59) and Schepel (2002: 44–47). A further view is offered at note 118 below.

[93] Case C-415/93 *ASBL v Jean-Marc Bosman, Royal Club Liégeois SA v Jean-Marc Bosman and Others and UEFA v Jean-Marc Bosman* [1995] ECR I-4921, para 147, 218–234, Central Marketing of the Commercial Rights to the UEFA Champions League [1999] OJ C 99/23, para 8, Joint Selling of the Commercial Rights of the UEFA Champions League [2003] OJ L 291/25, para 125–135. This approach is supported by Gyselen (2000: 439). Compare with case C-205/03 P *FENIN*, AG Opinion para 15.

[94] That the requirements are cumulative is suggested in Case C-244/94 *Fédération Française des Sociétés d'Assurance v Ministère de l'Agriculture et de la Pêche* [1995] ECR I-4013, para 18–20; that the requirements are alternative is suggested in Joined Cases C-264/01, C-306/01, C-354/01 and C-355/01 *AOK Bundesverband v Ichthyol-Gesellschaft Cordes* [2003] ECR I-2493, para 56, as the sickness funds are able to set the price and thus engaged in competition, but unable to determine-the level of service provided. See Case C-205/03 P *FENIN* AG Opinion, para 10 and Drijber (2005: 529–530).

(a) Compulsion Participation in the redistributive scheme must be compulsory.[95] In *Sodemare*, Advocate General Fennelly, considering solidarity (redistribution), felt that '[t]he system of compulsory contributions was indispensable to the principle of solidarity'.[96] The situation is then more appropriately seen as service provision funded from general taxation.[97] Since compulsory participation is essential, it is no surprise that provision of an *optional* supplementary old-age insurance scheme for self-employed farmers managed by a non-profit making organization is to be considered economic activity. Since 'membership of the . . . scheme *is optional* . . . a farmer wishing to supplement his basic pension will opt . . . for the solution which guarantees the better investment.'[98]

(b) No control over the cost of production Control over factors affecting the cost of production gives the potential to make profit.[99] In some activity identified as redistributive, the offeror of the goods or services has been unable to determine the level of service, or the quality/quantity of the good offered. Instead, this has been determined by the state.[100] It is not clear whether redistribution is only recognized when key factors affecting the cost of provision are determined by the state or whether other reasons putting factors affecting the cost of provision beyond the control of the entity can be recognized.

(c) No control over the price Control over the price at which the output is sold gives the potential to make profit.[101] In some activity identified as redistributive, the offeror of the goods or services has had no control over

[95] Joined Cases C-264/01, C-306/01, C-354/01 and C-355/01 *AOK Bundesverband v Ichthyol-Gesellschaft Cordes* [2003] ECR I-2493, para 50, Case C-475/99 *Ambulanz Glöckner v Landkreis Südwestpfalz* [2001] ECR I-8089, para 14, 20, Case C-70/95 *Sodemare SA v Regione Lombardia* [1997] ECR I-3395, AG Opinion para 29, Hervey (2000: 36), Menendez (2003: 378–379) and Turner (1969: 1208–1211).

[96] Case C-70/95 *Sodemare SA v Regione Lombardia* [1997] ECR I-3395, AG Opinion para 25.

[97] Compare with Case C-364/92 *SAT Fluggesellschaft mbH v Eurocontrol* [1994] ECR I-43, AG Opinion para 14.

[98] Case C-244/94 *Fédération Française des Sociétés d'Assurance v Ministère de l'Agriculture et de la Pêche* [1995] ECR I-4013, para 17, emphasis added.

[99] Case C-244/94 *Fédération Française des Sociétés d'Assurance v Ministère de l'Agriculture et de la Pêche* [1995] ECR I-4013, para 17, AG Opinion para 8, 17.

[100] Joined Cases C-264/01, C-306/01, C-354/01 and C-355/01 *AOK Bundesverband v Ichthyol-Gesellschaft Cordes* [2003] ECR I-2493, AG Opinion para 52 and Case C-218/00 *Cisal di Battistello Venanzio & Co. v Istituto Nazionale per L'assicurazione Contro Gli Infortuni Sul Lavoro (INAIL)* [2002] ECR I-691, para 71.

[101] Case C-244/94 *Fédération Française des Sociétés d'Assurance v Ministère de l'Agriculture et de la Pêche* [1995] ECR I-4013, AG Opinion at para 21.

the price at which the goods or services can be sold. Instead, the price has been determined by the state.[102] In *Diego Calì*, Advocate General Cosmas felt an important factor was 'the extent to which the entity whose activities are under review... has the power to influence the level of the consideration demanded in return for the services provided to users'.[103] Again, it is unclear whether redistribution is only recognized when the price is determined by the state or whether other reasons putting price beyond the control of the entity can be recognized.

(d) Absence of link between costs and price In *Cisal*, Advocate General Jacobs considered that an essential element of economic activity is that 'contributions and benefits are linked'.[104] For activity to be considered non-economic there must be no link between cost of provision and the price charged.[105] In *Albany* and *Poucet and Pistre*, profit could not be made as pension benefits were unrelated to pension contributions.[106] In *Cisal*, profit could not be made as insurance benefits were paid irrespective of fault and even if insurance contributions had not been paid.[107] In *Eurocontrol* profits could not be made as air traffic navigation services had to be provided even when customers had not paid and would not pay for the services.[108] The

[102] Joined Cases C-264/01, C-306/01, C-354/01 and C-355/01 *AOK Bundesverband v Ichthyol-Gesellschaft Cordes* [2003] ECR I-2493, para 56, Case C-218/00 *Cisal di Battistello Venanzio & Co. v Istituto Nazionale per L'assicurazione Contro Gli Infortuni Sul Lavoro (INAIL)* [2002] ECR I-691, para 39–40, 44–45, AG Opinion para 25, 76, Joined Cases C-159/91 and C-160/91 *Christian Poucet v Assurances Generales de France (AGF) and Caisse Mutuelle Regionale Du Languedoc-Roussillon (CAMULRAC) and Daniel Pistre v Caisse Autonome Nationale de Compensation de l'Assurance Vieillesse des Artisans (CANCAVA)* [1993] ECR I-637, para 11, and Case C-364/92 *SAT Fluggesellschaft mbH v Eurocontrol* [1994] ECR I-43, para 29.

[103] Case C-343/95 *Diego Cali & Figli Srl v Servizi Ecologici Porto di Genova SpA (SEPG)* [1997] ECR I-1547, AG Opinion para 42. Also Joined Cases C-159/91 and C-160/91 *Christian Poucet v Assurances Generales de France (AGF) and Caisse Mutuelle Regionale Du Languedoc-Roussillon (CAMULRAC) and Daniel Pistre v Caisse Autonome Nationale de Compensation de l'Assurance Vieillesse des Artisans (CANCAVA)* [1993] ECR I-637, para 15 and Lasok (2004: 384–385).

[104] Case C-218/00 *Cisal di Battistello Venanzio & Co. v Istituto Nazionale per L'assicurazione Contro Gli Infortuni Sul Lavoro (INAIL)* [2002] ECR I-691, AG Opinion para 62, also para 80.

[105] Case C-218/00 *Cisal di Battistello Venanzio & Co. v Istituto Nazionale per L'assicurazione Contro Gli Infortuni Sul Lavoro (INAIL)* [2002] ECR I-691, AG Opinion para 81.

[106] Case C-67/96 *Albany International BV v Stichting Bedrijfspensioenfonds Textielindustrie* [1999] ECR I-5751, para 78–79, AG Opinion para 342, and Joined Cases C-159/91 and C-160/91 *Christian Poucet v Assurances Generales de France (AGF) and Caisse Mutuelle Regionale Du Languedoc-Roussillon (CAMULRAC) and Daniel Pistre v Caisse Autonome Nationale de Compensation de l'Assurance Vieillesse des Artisans (CANCAVA)* [1993] ECR I-637, para 10, 15.

[107] Case C-218/00 *Cisal di Battistello Venanzio & Co. v Istituto Nazionale per L'assicurazione Contro Gli Infortuni Sul Lavoro (INAIL)* [2002] ECR I-691, para 35–36, 44–45, AG Opinion para 68–76.

[108] Case C-364/92 *SAT Fluggesellschaft mbH v Eurocontrol* [1994] ECR I-43, AG Opinion para 13.

absence of a link excludes the possibility that profit can be made. Instead, subsidization is required.[109]

2. Public goods

Public goods and services possess two characteristics that make profit impossible.[110] The first characteristic is that such goods are non-rivalrous in consumption.[111] This means that, once produced, an infinite number of consumers can enjoy the good without increased production cost or diminished enjoyment by other consumers. National defence is typically used as an example.[112] The cost of defending the territory of the United Kingdom does not increase when an additional member is added to the population; the existing population does not become more vulnerable because there is an additional resident to defend. The second characteristic is that the benefits are non-excludable.[113] It is impossible to prevent people from enjoying the benefits once the good is produced. Again, the example of national defence can be used. All the residents benefit once the territory of the United Kingdom is protected; it is impossible to defend the territory of the United Kingdom without defending all the residents. The inability to exclude people from the benefits of consumption removes the incentive for individuals to pay for consumption: individuals benefit regardless of whether they pay for the good or service, so why pay?[114] The problem then is that private producers have no incentive to produce the goods because it is impossible to profit from a good which non-payers cannot be prevented from enjoying.[115] Government making payment compulsory, usually through a system of general taxation, is one way to solve the problem.[116]

The Court appears to have recognized that it is not possible to profit from the provision of public goods.[117] In *Poucet and Pistre*, Advocate General Tesauro considered that activity is not economic when the task

[109] On the need for subsidy to fund the European social model see Hatzopoulos (2002: 684) and Case 1006/2/1/01 *Bettercare Group Limited v the Director General of Fair Trading* [2002] CAT 7, para 231–235.

[110] Stiglitz and Driffill (2000: 126) point out that few goods are *pure* public goods, but instead exhibit the characteristics to a greater or lesser extent.

[111] Stiglitz (2000: 128–129), Stiglitz and Driffill (2000: 124), Carlton and Perloff (2000: 82), Case and Fair (1999: 387–388).

[112] Hummel (1990).

[113] Stiglitz (2000: 128–129), Stiglitz and Driffill (2000: 124), Case and Fair (1999: 388).

[114] Stiglitz (2000: 130–146).

[115] This is a classic example of the free-rider problem: Groves and Ledyard (1977), Case and Fair (1999: 392).

[116] Stiglitz and Driffill (2000: 126–127), Stiglitz (2000: 129–135), and Case and Fair (1999: 289).

[117] Buendia Sierra (1999: para 1.158–1.159, 1.188–1.190, 1.198–1.201).

'*can only* be performed by or on behalf of a public body'.[118] The recognition that public goods cannot be provided in the market place is evident in *Albany*, Advocate General Jacobs writing that 'it seems to follow from paragraph 22 of the judgment in *Höfner* that the competition rules do not apply if the activity in question has always been and is *necessarily carried out by public entities*'.[119]

Both *Eurocontrol* and *Diego Calì* show recognition that effective provision of a public good is impossible absent the coercive power of the state.[120] In *Eurocontrol*, the Court accepts air traffic control as a public good.[121] The service is non-excludable: for the service to function effectively it is necessary to provide the service to airlines irrespective of whether they have paid. Advocate General Tesauro noted that:

> the fact we are dealing with a service, not in the economic sense and provided principally for businesses (airline companies), but aimed at the community as a whole, seems to me to be confirmed by the observation made during the hearing . . . that control is exercised in respect of any aircraft, within the air space under the authority of Eurocontrol, irrespective of whether or not the owner has paid the route charges.[122]

Payment to the service provider can only be assured by state compulsion, and this again is recognized by the Advocate General writing that 'those charges undoubtedly constitute a tax burden, since they are a sort of financial contribution to the costs incurred by the States, payable by the individual for the benefits he has received'.[123] The Court thus accepted that air-traffic control activities are 'typically those of a public authority. They are not of an economic nature justifying the application of the Treaty rules of competition'.[124]

[118] Joined Cases C-159/91 and C-160/91 *Christian Poucet v Assurances Générales de France (AGF) and Caisse Mutuelle Regionale du Languedoc-Roussillon (CAMULRAC) and Daniel Pistre v Caisse Autonome Nationale de Compensation de l'Assurance Vieillesse des Artisans (CANCAVA)* [1993] ECR I-637, AG Opinion para 12, emphasis added. The AG is actually speaking about solidarity, raising the question of whether redistributive activity is not considered economic because it is a public good. Compare with Prosser (2005: 35–37).

[119] Case C-67/96 *Albany International BV v Stichting Bedrijfspensioenfonds Textielindustrie* [1999] ECR I-5751, AG Opinion para 314, emphasis added. Also Case C-475/99 *Ambulanz Glöckner v Landkreis Südwestpfalz* [2001] ECR I-8089, para 20 and Buendia Sierra (1999: para 1.158).

[120] See discussion of imperium at text accompanying notes 55–56 above.

[121] Case C-364/92 *SAT Fluggesellschaft mbH v Eurocontrol* [1994] ECR I-43, para 16.

[122] Case C-364/92 *SAT Fluggesellschaft mbH v Eurocontrol* [1994] ECR I-43, AG Opinion para 13, emphasis added, judgment para 25 and Buendia Sierra (1999: para 1.166–1.170).

[123] Case C-364/92 *SAT Fluggesellschaft mbH v Eurocontrol* [1994] ECR I-43, AG Opinion para 14.

[124] Case C-364/92 *SAT Fluggesellschaft mbH v Eurocontrol* [1994] ECR I-43, para 30, also AG Opinion para 12–13.

Various functions of port authorities have been classified as *economic*. For example, the loading and unloading of ordinary freight and piloting services in the Port of Genoa are seen as economic activity.[125] However, in *Diego Calì*, maintaining a clean environment at the port is recognized as a public good. The protection of the environment at the oil port of Genoa-Multedo was the responsibility of Consorzio Autonomo del Porto (CAP), a public body.[126] CAP delegated the task to Servizi Ecologici Porto di Genova (SEPG), a private entity, to monitor and enforce safety procedures designed to prevent oil spillages and thus protect the environment. SEPG sent Diego Calì a bill for approximately €4,500 in relation to the monitoring functions performed. Diego Calì refused to pay, objecting that they had not requested or received any service from SEPG. When ordered to pay, Diego Calì argued that the system of pollution control was in breach of competition law, raising the question of whether SEPG was engaged in economic activity and thus operating as an undertaking.

SEPG's task is characterized as ensuring the provision of the non-excludable benefits of a clean environment, Advocate General Cosmas attaching importance to the fact that:

> the fundamental aim of SEPG's anti-pollution activity is not only to guarantee the safety... of those districts of Genoa close to the port where tourism is a growth industry, but also to protect the port environment and, in the final analysis, ensure that public assets are properly preserved.[127]

The Court considered that the nature of the activity, 'protection of the environment in maritime areas', constituted one of 'the essential functions of the State'.[128] This makes it clear that SEPG is considered to be providing a public good, forcefully confirmed by Advocate General Cosmas writing that 'the maritime zone of the Porto Petroli, that is to say a public asset, is being protected in the interest of the State and of citizens'.[129]

[125] Case C-179/90 *Merci v Siderurgfica* [1991] ECR I-5889 and Case C-18/93 *Corsica Ferries Italia v Corpo dei Piloti del Porto di Genova* [1994] ECR I-1783.

[126] Information on Genoa port services is available from http://www.informare.it/news/forum/capoc1uk.htm.

[127] Case C-343/95 *Diego Cali & Figli Srl v Servizi Ecologici Porto di Genova SpA (SEPG)* [1997] ECR I-1547, AG Opinion para 44.

[128] Case C-343/95 *Diego Cali & Figli Srl v Servizi Ecologici Porto di Genova SpA (SEPG)* [1997] ECR I-1547, para 22, AG Opinion para 29, 43–64.

[129] Case C-343/95 *Diego Cali & Figli Srl v Servizi Ecologici Porto di Genova SpA (SEPG)* [1997] ECR I-1547, AG Opinion para 46.

In order to pay for the provision of a clean environment, a non-excludable good, a compulsory charge was levied on each vessel using the port, based on the size of the vessel, the quantity of oil being transported, and type of service SEPG performed.[130] The ability to levy the compulsory charge derives from the taxation powers of the CAP, a public body exercizing imperium.[131] Consequently, the activity 'is not of an economic nature justifying the application of the Treaty rules on competition'.[132] Both the recognition of a public good and the influence of such recognition on the characterization of the nature of the activity are captured in the Opinion of Advocate General Cosmas, who concluded that:

> the activity of SEPG cannot conceivably be carried out within a competitive system, since that would jeopardize, if not destroy, the effectiveness of the system of safeguards as regards both the port environment and the safety of port users and inhabitants of the surrounding areas. It is therefore a public service unrelated to commercial profit-making activity. Furthermore, that this service is provided for the benefit of the whole of the community is also apparent from the fact that the surveillance has to be exercised regardless whether the fees owed by any particular vessel have been paid.[133]

IV. Assessing the functional approach

Two claims have been made: first, the Court takes a functional approach in determining the addressees of Community competition law. We are concerned with activity rather than with the institutions carrying out the activity. Second, the activity we are concerned with is economic activity. In relation to the second claim, it is further argued that three cumulative elements (the offer of goods or services, the bearing of risk, and the potential to make profits) enable economic activity to be identified. This next section takes the analysis further by considering why a functional approach is taken and asking whether economic activity is the only activity that *is* and *ought* to be subject to the Community competition law rules.

[130] Case C-343/95 *Diego Cali & Figli Srl v Servizi Ecologici Porto di Genova SpA (SEPG)* [1997] ECR I-1547, para 8, 13, 24, AG Opinion para 52–54.

[131] Case C-343/95 *Diego Cali & Figli Srl v Servizi Ecologici Porto di Genova SpA (SEPG)* [1997] ECR I-1547, AG Opinion para 52–54.

[132] Case C-343/95 *Diego Cali & Figli Srl v Servizi Ecologici Porto di Genova SpA (SEPG)* [1997] ECR I-1547, para 23.

[133] Case C-343/95 *Diego Cali & Figli Srl v Servizi Ecologici Porto di Genova SpA (SEPG)* [1997] ECR I-1547, AG Opinion para 49.

A. *The need for a functional approach*

We should recall Deringer's preference for an institutional approach; that undertaking 'cannot be the activity . . . but only a subject in relation to a certain doing'.[134] Why then is a functional approach required? The need for a functional approach can be explained in three stages. First, it is argued that the EC Treaty incorporates a public/private divide: there are rules applicable to private parties and rules applicable to the Member States.[135] Second, it is argued that difficulties in distinguishing the public from the private sphere exist. Finally, it is argued that the functional approach is used to distinguish the public from the private sphere. Knowledge of a public/private divide in the EC Treaty and of the necessity, difficulty, and consequences of maintaining its existence are central and essential to understanding the task the Court is engaged in when determining the meaning of undertaking.

1. *The public/private divide in the EC Treaty*

It is important at the outset to consider why the state should be afforded different treatment under law than the citizen.[136] Differential treatment of public authorities and private (corporate) citizens can be justified to reflect the differing roles played by citizen and the state in a constitutional democracy. Essentially, the state is charged with carrying out certain tasks that only the state can perform.[137] The state is endowed with special powers to enable it to perform these tasks: the powers are exercised in a representative capacity.[138] The tasks in which the state engages, the power available in order to achieve these tasks, and the way in which the power arises, justify differential treatment. Public power has a democratic legitimacy; its exercise has a degree of political accountability; and the existence and use are justified by a public interest.[139] The underlying assumption is that public power is exercised in the public interest, and the public interest is determined through democratic representation.

When the exercise of public power conflicts with legally protected public interests, there is a conflict between competing public interests—one

[134] Deringer (1968: 5). [135] van den Bogaert (2002: 123).

[136] Oliver (1999), Harlow and Rawlings (1997: 7–9), Aronson (1997: 52–56), Galbraith (2004: 47–51, 69–72), and Kennedy (1982: 1349).

[137] Bamforth (1997: 137, 140–151), Loughlin (2003: 1, 7–12), and Harlow and Rawlings (1997: 9–15).

[138] Bamforth (1997: 138–140, 151–154), Loughlin (2003: 46–47, 13–14, 54–80), and Craig (2001: 128).

[139] Case C-205/03 P *FENIN*, AG Opinion para 26, Oliver (1999: 12–13), and Stiglitz (2000: 13–14).

deriving legitimacy from law, the other deriving legitimacy from democratic representation.[140] The exercise of *public* power in a manner infringing a *public* interest should be treated differently from the exercise of *private* power infringing the same *public* interest: private actors lack the same democratic legitimacy.[141] The exercise of *public* power needs distinctive substantive and procedural rules that take account of the nature of the activity engaged in, the legitimacy of those activities, and the appropriate allocation of tasks between the courts and the legislature.[142]

There has been much work on whether a public/private divide exists in the EC Treaty.[143] The early approach to Treaty interpretation identified rules imposing obligations on private entities and rules imposing obligations on Member States: *inter alia*, the free-movement rules (Articles 28, 39, 43, 49/50 EC) applied to Member State activity (EC public law) and Articles 81 and 82 EC applied to the activity of private entities (EC private law).[144] Taking the Treaty as a coherent whole, all practices preventing the attainment of Community goals are prohibited, but there are specific provisions to deal with restrictions caused by private parties and specific provisions to deal with restrictions caused by the Member States: the same principles apply within the public and the private sphere, but the specific rules differ.[145]

2. *Challenges to the public/private divide*

Whilst the public/private divide can be identified from the Treaty text and its early interpretation, two factors challenge the maintenance of the divide. First, the public/private divide presupposed that those operating in each sphere could be identified, and identified *institutionally*. States are thought of as acting to safeguard *non-economic* interests and are the addressees of the free movement provisions; private parties are thought of as acting to

[140] Public law must then deal with competing public interests: Maduro (1997: 73–79).

[141] Cruz (2002: 88–89, 111, 132–134, 149, 155–161).

[142] Case C-205/03 P *FENIN*, AG Opinion para 26-27, Oliver (1999: 19–22, 80–88), Loughlin (2003: 5), Maduro (1997: 55–56), Snell (2003: 324, 337), and Alder (1997: 166–167).

[143] The debate over a public/private divide is advanced in the terminology of whether particular provisions have horizontal or vertical direct effect. On this debate see: Pescatore (1987), Marenco (1987), Milner-Moore, Cruz (2002), Quinn and MacGowan (1987), van den Bogaert (2002), Oliver (1999: 117–122) and Snell (2002).

[144] van den Bogaert (2002: 123), Mortelmans (2001: 622–623, 635–636), Hatzopoulos (2000: 76–80), Weatherill and Beaumont (1999: 521–522), O'Loughlin (2003: 62), Snyder (1990: 92–93), Slot (1987: 181), Bamforth (1997: 143–151), and Whish (2003: 211).

[145] Weatherill (1989: 87–92) and Weatherill (1996: 1003) arguing the Treaty should be seen as a coherent whole.

safeguard *economic* interests and are the addressees of the competition provisions.[146] The distinction blurs when what can be seen as public functions (non-economic activities) are carried out by private entities.[147] A process of privatization and contracting out during the 1980s transferred tasks previously performed by the state to private firms operating in the market place.[148] Shorn of its mystique it is also plain that sometimes the state is simply pursuing private economic interests.[149] It can no longer be said that the state *as an institution* pursues non-economic public interests or that private entities *as institutions* pursue private economic interests. Determining the addressee of the particular rules is then by necessity a *functional* rather than institutional enterprise.[150] 'The enterprise seeks to capture the intuition that some activities are simply the type of things done by governments, or that might have been done by government instead.'[151] This means that sometimes the activities of a private entity will be considered under rules applicable to the public sphere because the nature of the activity is seen as public.[152] Similarly, sometimes the activities of the state will be considered under the rules applicable to the private sphere because the nature of the activity is seen as private.[153]

A second challenge is that the divide seemed to presume that private parties, though not subject to EC public law, 'are already adequately provided for by other articles of the Treaty, in particular those dealing with the competition rules'.[154] However, some activities in the private sphere fall outside the EC private law rules but infringe principles supposedly protected by the Treaty. Two examples are unilateral conduct by non-dominant undertakings and action by non-undertakings that affects trade

[146] Mortelmans (2001: 635–636).

[147] See Sections III.A.2 and III.C.1 and III.C.2 above.

[148] Estrin (1994), Wright (1994), Wright and Perrotti (2000), and Prosser (2005: 20–24, 44–47). On the transfer of executive functions to the market place: Harlow and Rawlings (1997: 15–23), Oliver (1999: 11–12), Allison (1997: 79–83), Aronson (1997: 40–45), Freedland (1994), Mortelmans (2001: 640), Hertz (2002), and Monbiot (2000).

[149] Stiglitz (2000: 28–29, 136–141), Szyszczak (2004: 189–190), and Section III.C above.

[150] Mortelmans (2001: 624, 634–635) and Woolf (1995: 63–64).

[151] Milner-Moore (text accompanying note 125). On the attempt to distinguish state from non-state activity see Milner-Moore (at text accompanying notes 110–125), Pescatore (1987: 407–416), Schepel (2002), Allison (1997: 84–88), Alder (1997: 166–170), and Oliver (2000).

[152] Case 249/1981 *Commission v Ireland* [1982] ECR 4005.

[153] Case 118/85 *Commission v Italian Republic* [1987] ECR 2599.

[154] Quinn and MacGowan (1987: 166). Milner-Moore (text accompanying note 83) and Gormley (2002) reject the idea that the competition rules can achieve the same outcome as the free-movement rules.

between Member States.[155] And some activities in the public sphere fall outside the EC public law rules but again infringe principles supposedly protected by the Treaty. For example, it is possible for the state to fix prices or create circumstances that render the competition rules superfluous.[156] There are gaps in the Treaty. The question then is whether to close the lacuna by applying both sets of Treaty rules to all types of practices (removing the public/private divide) or whether the lacuna is deliberate.[157] In *Bosman* Advocate General Lenz took the former view, writing that the competition and free-movement rules could be applied simultaneously since '[n]o reason can be seen why the rules at issue in this case should not be subject both to Article [39] and to EC competition law . . . so that in principle both sets of rules may be applicable to a single factual situation'.[158] This view reaches its zenith in *Angonese*, when the Court finds that Article 39 EC is capable of binding private parties as well as Member States.[159] The current jurisprudence suggests the absence of a divide when the issue concerns persons, but the existence of a divide when goods and services are concerned.[160]

Gap closing is not without problems. Whilst the free-movement rules apply when there is an effect on intra-Community trade, the competition rules additionally require a restriction on competition or an abuse of a dominant position (to the extent that abuse of dominance differs from

[155] Case T-41/96 *Bayer AG v Commission* [2000] ECR II-3383, para 176–181 and Joined Cases C-2/01 P and C-3/01 P *Bundesverband der Arzneimittel-Importeure eV and Commission v Bayer AG* [2004] ECR I-23, para 69–70, 101 seem to confirm the existence of the gap and rule out certain interpretative means of filling the gap. Also Snell (2002: 228–236).

[156] Case 13/77 *GB-INNO-BM v ATAB* [1977] ECR 2115, Case 229/83 *Leclerc* [1985] ECR 1, Case 231/83 *Henri Cullet and Chambre Syndicale Des Reparateurs Automobiles Et Detaillants De Produits Petroliers v Centre Leclerc à Toulouse and Centre Leclerc à Saint-Orens-De-Gameville* [1985] ECR 305, Case T-41/96 *Bayer AG v Commission* [2000] ECR II-3383, para 56, Faull and Nikpay (1999: para 5.01–5.07), Whish (2003: 211–217), Gyselen (1994), and Chung (1995).

[157] Pescatore (1987), Marenco (1987), and Quinn and MacGowan (1987). For accounts of various gap closing techniques and the implications of their adoption: Quinn and MacGowan (1987), Snell (2002: 228–243, Cruz (2002: 85–163) and Weatherill (1989: 80–87).

[158] Case C-415/93 *ASBL v Jean-Marc Bosman, Royal Club Liégeois SA v Jean-Marc Bosman and Others and UEFA v Jean-Marc Bosman* [1995] ECR I-4921, AG Opinion para 253. The Court decided not to deal with the issue at para 138: Weatherill (1989: 60), Weatherill (1996: 1000–1003, 1018–1026), and van den Bogaert (2000: 557).

[159] Case C-281/98 *Roman Angonese* [2000] ECR I-4139. For a summary of the extent to which the free-movement rules are capable of binding private parties: van den Bogaert (2002: 132–133). On the subjecting of the 'public sector' to market rules see Spaventa (2004a) and Hervey (2000). The inapplicability of the competition rules when the free-movement rules apply is seen as paradoxical to Belhaj and van de Dronden (2004: 686).

[160] van den Bogaert (2002: 149–150) and Spaventa (2004b).

a restriction of competition). Subjecting activities to the simultaneous application of the competition and free-movement rules would render the additional requirements in the competition rules redundant.[161] Closing gaps by the simultaneous application of the competition and free-movement rules would distort what some see as an intended relationship between rules in the two spheres.[162] As Advocate General Capotorti noted, '[t]here is a distinction between Articles [28] and [29] on the one hand and Articles [81] and [82] on the other, not only with regard to those subject to the prohibitions but also with regard to the nature of the behaviour which is prohibited'.[163]

3. *The functional approach as a means of drawing the public/private divide*

The presumption underpinning the need for different rules for public and private spheres is that private actors pursue self-interest whilst public actors promote the general public interest.[164] Competition is protected in the public interest and Article 81 EC operates to ensure that those pursuing a *private* interest do not harm the *public* interest.[165] Even when a distinction is justified, 'it must be possible to make the distinction'.[166] It is clear that an *institutional* approach can no longer be used to identify those pursuing a private interest.[167] The functional approach provides a means of allocating functions to the public or the private sphere so that the appropriate Treaty rules can be applied.[168] This sphere allocation function can be seen in

[161] Quinn and MacGowan (1987: 167–170), van den Bogaert (2002: 140), and Weatherill (1989: 90–92). Redundancy is rejected by Milner-Moore (at text accompanying notes 286–292).

[162] Weatherill (1989: 88–92) suggests the maintenance of the divide creates coherent but less effective Community law.

[163] Case 82/77 *Openbaar Ministerie of the Kingdom of the Netherlands v Jacobus Philippus Van Tiggele* [1978] ECR 25, 47. See van den Bogaert (2002: 139–143) and Slot (1987: 187).

[164] Case C-2/1991 *Criminal Proceedings against Wolf W Meng* [1993] ECR I-5751, 5770, Joined Cases T-528/1993, T-542/1993, T-543/1993 and T-546/1993 *Metropole Télévision SA v Commission (Eurovision)* [1996] ECR II-649, para 117–118, Case C-67/96 *Albany International BV v Stichting Bedrijfspensioenfonds Textielindustrie* [1999] ECR I-5751, AG Opinion para 184 and Section IV.A.2 above. Compare with Prosser (2005).

[165] Case C-67/96 *Albany International BV v Stichting Bedrijfspensioenfonds Textielindustrie* [1999] ECR I-5751, AG Opinion para 184, 206, Case C-2/1991 *Criminal Proceedings against Wolf W Meng* [1993] ECR I-5751, 5770, Joined Cases T-528/1993, T-542/1993, T-543/1993 and T-546/1993 *Metropole Télévision SA v Commission (Eurovision)* [1996] ECR II-649, para 117–118, and Willimsky (1997: 54). This is forcefully confirmed in US competition law by *National Society of Professional Engineers v United States* 435 US 679, 682–692 (1978), discussed in American Bar Association (Antitrust Section) and Hartley (chairman) (1999: 116–119) and Pitofsky (1995).

[166] Kennedy (1982: 1349). [167] See Section IV.A.2 above.

[168] Joined Cases C-264/01, C-306/01, C-354/01 and C-355/01 *AOK Bundesverband v Ichthyol-Gesellschaft Cordes* [2003] ECR I-2493, AG Opinion para 52, citing Case C-185/91 *Reiff* [1993]

Wouters, the Court drawing attention to the question of whether an entity 'is to be treated as an association of *undertakings* or, on the contrary, as a public authority'.[169]

The functional approach set out in *Höfner* identifies the occupants of the private sphere as self-interested and uses the proxy of *homo economicus* to populate the sphere: economic actors are characterized as rational maximizers of *self*-interest.[170] When an entity has the potential to make profit by taking the financial risk of offering goods or services to the market, they are engaged in economic activity and can be *presumed* to pursue private self-interest.[171] The presumption that *homo economicus* is a rational maximizer of self-interest justifies the application of the substantive rules contained in EC competition law, which are applied to ensure that those pursuing their *private* self-interest do not restrict competition, which is protected in the *public* interest.[172] Articles 81 and 82 EC are aspects of private law and the function of economic activity is used to define the scope of the private sphere.[173] If, despite the presumption, an entity engaged in economic activity pursues the public interest this is considered at the stage of justification: Article 86(2) EC is one tool enabling such justification.[174]

Entities that do not satisfy the conditions for economic activity engage in acts 'contrary to the pecuniary interests of the parties concerned and thus do not pose the dangers normally associated with collaboration among competitors'.[175] For example, the presumption that economic agents pursue private self-interest is easily rebutted when the entity is under an

ECR I-5801, para 16, Case C-67/96 *Albany International BV v Stichting Bedrijfspensioenfonds Textielindustrie* [1999] ECR I-5751, AG Opinion para 184, 206 and Cruz (2002: 156). On the attempted resolution of the problem in UK public law: Hunt (1997: 27–33).

[169] Case C-309/99 *Wouters v Algemene Raad Van de Nederlandse Orde Van Advocaten (Raad Van de Balies Van de Europese Gemeenschap, Intervening)* [2002] ECR I-1577, para 56, emphasis added. Also AG opinion para 70, Joined Cases C-180/98 to C-184/98 *Pavel Pavlov v Stichting Pensioenfonds Medische Specialisten* [2000] ECR I-6451, para 87–88, and Cruz (2002: 148–149, 154).

[170] Mitchell (2002: 1941, note 58), Persky (1995), Bowles and Gintis (1993), Hausman and McPherson (1996: 27–30), Zafirovski (2000), Daly and Cobb (1990: 5), and Veljanovski (1982: 27–31).

[171] See Section III above.

[172] See references cited at note 165 above.

[173] Whish (2003: 211), considering that Articles 81 and 82 EC are private law. For arguments that some bodies must be considered as occupants of the public sphere institutionally: Oliver (2000: 477–479). For a discussion of problems caused by the choice of functional over an institutional approach: Cruz (2002: 85–87) and Szyszczak (2004: 194–199).

[174] Faull and Nikpay (1999: para 5.122–5.155).

[175] Turner (1969: 1210). Compare with Case C-309/99 *Wouters v Algemene Raad Van de Nederlandse Orde Van Advocaten (Raad Van de Balies Van de Europese Gemeenschap, Intervening)* [2002] ECR I-1577, para 56 and Drijber (2005: 530).

obligation to pursue the public interest: the entity then falls outside the scope of Community competition law on this basis.[176] The activities of those not seen as pursuing self-interest (which this chapter views as synonymous with economic/private activity) can be presumed as the pursuit of a public interest and thus more appropriately subject to the rules more able to take account of competing public interests being pursued.[177]

B. *Economic activity as the sole function*

Article 81 EC applies to undertakings; undertakings are engaged in economic activity; it is possible to distinguish economic from non-economic activity. The functional approach established in *Höfner* thus suggests that Article 81 EC applies to economic activity and *does not* apply to non-economic activity. This raises a number of questions: what rules apply to non-economic activity and how do these rules differ from those that apply to economic activity? The response of this chapter is that the free-movement rules apply to non-economic activity; the rules differ because the competition rules protect the public interest from private interests whilst the free-movement rules balance competing public interests. This response assumes that the free-movement rules are capable of balancing competing public interests; that only the free-movement rules are capable or appropriate for balancing competing public interests; and that this is represented in a public/private divide that (1) exists within the Treaty; (2) can be identified and operationalized; (3) and is desirable.

At all three levels there is room for disagreement. Article 81 EC challenges the idea that the competition rules *do not* apply and *cannot* apply to non-economic activity. This is because in addition to undertakings, Article 81 EC also applies to associations of undertakings. If the functional definition of undertaking given in *Höfner* captures all economic activity

[176] Joined Cases C-264/01, C-306/01, C-354/01 and C-355/01 *AOK Bundesverband v Ichthyol-Gesellschaft Cordes* [2003] ECR I-2493, AG Opinion para 53, Case C-309/99 *Wouters v Algemene Raad Van de Nederlandse Orde Van Advocaten (Raad Van de Balies Van de Europese Gemeenschap, Intervening)* [2002] ECR I-1577, AG Opinion para 70, Joined Cases C-180/98 to C-184/98 *Pavel Pavlov v Stichting Pensioenfonds Medische Specialisten* [2000] ECR I-6451, para 87, Case 249/1981 *Commission v Ireland* [1982] ECR 4005, para 15, Case 222/1982 *Apple and Pear Development Council v K J Lewis Ltd* [1983] ECR 4083, para 2–7, 17, Snell (2002: 218–221), Cruz (2002: 148–149, 154), and Oliver (2000: 482–484).

[177] On the inability of the competition rules to consider competing public interests internally: Marenco (1987: 421–423), Turner (1969: 1210), and Case C-2/1991 *Criminal Proceedings against Wolf W Meng* [1993] ECR I-5751, 5770. On the ability of EC public law to deal with competing public interests internally: Maduro (1997: 73–79).

then associations of undertakings must be addressed when engaged in non-economic activity, otherwise the association would be an undertaking in its own right and 'associations of undertakings' otiose. A number of interesting questions then emerge: why and how does Community competition law scrutinize non-economic activity; is the substantive assessment that occurs in relation to non-economic activity the same as the substantive assessment that occurs in relation to economic activity; what non-economic activity is appropriately scrutinized; why is it only the non-economic activity of associations of undertakings that falls with the scope of Article 81 EC; why is Article 82 EC only addressed to undertakings and not associations of undertakings?

A number of these challenges are present in *Wouters*, a decision in which the Court considered a *regulatory* rule preventing lawyers from entering into partnership with accountants.[178] It is clear that the activity was regulatory and that regulation is not economic activity.[179] The rule came within the scope of Article 81 EC, not because it was economic activity, but because it was promulgated by *an association of undertakings*.[180] Once within Article 81 EC the challenge is to find an appropriate standard by which to assess non-economic, regulatory, activity.[181] The standard the Court applies seems to be that which applies under the free-movement rules; the Court applies free-movement principles rather than competition principles.[182] Non-economic activity is subject to public rather than private law and the Court should be mindful not to capture non-economic activity when using the association of undertakings limb. The term would still play an independent role for the purpose of attribution, so that *Faull & Nikpay* write that 'the importance of the concept "decisions by associations of undertakings" therefore lies in the fact that it enables those applying Article 81(1) to hold associations liable for the anti-competitive behaviour of their members'.[183]

The problem with *Wouters* is that what occurs is so opaque. The standard of assessment for the activity, captured by less than stringent application of the functional approach, was clearly to be found in public rather than private law. A clearer, though by no means clear, discussion of whether conduct

[178] The national legal framework is more fully set out in Deards (2002: 619).

[179] See Section III.A.3.

[180] Case C-309/99 *Wouters v Algemene Raad Van de Nederlandse Orde Van Advocaten (Raad Van de Balies Van de Europese Gemeenschap, Intervening)* [2002] ECR I-1577, para 71, AG Opinion at para 56–87.

[181] See note 44 above.

[182] Deckert (2000: 176) and Monti (2002: 1089–1090).

[183] Faull and Nikpay (1999: para 2.44). Also Deringer (1968: para 118) and Whish (1993: 168–169).

should be subject to public as opposed to private law is found in the UK case of *Bettercare*.[184] North & West Belfast Health and Social Services Trust was under a statutory duty to provide residential care and nursing home services for which it received an annual governmental grant, and was also under a statutory duty to recover the full cost of providing the services from the recipient, subject to ability to pay.[185] The CAT asked what law was appropriate to apply in the situation. Since the application of competition law would interfere with the distribution by the state of funds between its social welfare priorities, the OFT argued that 'such activities are not suitable for control under competition law'.[186] Rather than competition law, private law, recourse should be had to public law or extra-legal methods:

> The principal alternative methods . . . [of] challenge . . . include direct approaches to the responsible Minister . . . . Other possibilities . . . include a complaint to the Northern Ireland Commissioner for Complaints, a complaint to the Northern Ireland Audit Office or, possibly, a claim for *judicial review*.[187]

The CAT was not convinced that '[i]f the Act does not apply . . . the independent provider would have any other adequate legal remedy. The largely unexplored area of judicial review . . . does not appear to us to be particularly promising'.[188] Regardless of the merits of the outcome of the case, it is clear that consideration is given to whether public or private law is the applicable meter of assessment.[189]

V. Conclusion

It has been argued that a functional approach is taken to determine the scope of Article 81 EC.[190] The function that is subject to Article 81 EC is

[184] Case 1006/2/1/01 *Bettercare Group Limited v the Director General of Fair Trading* [2002] CAT 7. See references and discussion at note 39 above.

[185] Case 1006/2/1/01 *Bettercare Group Limited v the Director General of Fair Trading* [2002] CAT 7, para 110–115, 136–137, 140–149, 187–188.

[186] Case 1006/2/1/01 *Bettercare Group Limited v the Director General of Fair Trading* [2002] CAT 7, para 60.

[187] Case 1006/2/1/01 *Bettercare Group Limited v the Director General of Fair Trading* [2002] CAT 7, para 61, emphasis added.

[188] Case 1006/2/1/01 *Bettercare Group Limited v the Director General of Fair Trading* [2002] CAT 7, para 261, citations omitted. On the developing ability to apply UK public law to private parties see Alder (1997), Aronson (1997: 45–51), Beloff (1989: 104–110), Borrie (1989), Craig (1991), Craig (1997), Hunt (1997: 27–39), and Oliver (1999: 88–92, 201–218).

[189] *Regina v Panel on Take-Overs and Mergers, ex p Datafin (Court of Appeal)* [1987] QB 815, 820, 839 shows that the debate is not new, though on that occasion public law rather than private law was applied.

[190] See Section II above.

economic activity. The Court suggests three cumulative elements of economic activity: the offer of goods or services; the bearing of economic or financial risk; and the potential to make profit.[191] By identifying the three positive elements of economic activity, it is possible to see a number of situations when the Court considers that the elements are not satisfied. The offer of goods or services is required: work, consumption, and regulation do *not* satisfy this requirement.[192] The potential to make profit must exist: it is impossible to profit from redistribution or the provision of public goods.[193] The elements of economic activity identified thus provide a framework in which it is possible to understand some confusing and complex case law. If the elements of economic activity identified in this chapter are those that must exist for the Court to find that an entity is an undertaking, it is a duty of the Court to state them more clearly than it has done thus far.[194] Further, in rejecting certain activity as non-economic, it is incumbent on the Court to explain which elements have not been satisfied, and to explain why those elements have not been satisfied. Thus far, the Court has failed to fulfil these duties and legal certainty has suffered as a result.[195]

In addition to identifying the functional approach and the elements of the function (i.e., the elements of economic activity), the need for a functional approach has been considered. It has been argued that the EC Treaty contains a public/private divide and that knowledge of the divide and of the necessity, difficulty, and consequences of maintaining its existence are central and essential to understanding the task the Court is engaged in when determining the addressees of the Community competition rules. Kennedy is of the view that a public/private divide should not be drawn unless 'it seems plain that situations should be treated differently'.[196] It has been argued that different treatment is justified by a presumption underlying rules of the private sphere that its occupants are self-interested and the presumption underlying rules of the public sphere that its occupants operate in pursuit of the public interest.[197] Kennedy is also of the view that a public/private divide should not be drawn unless it is 'possible to make the distinction'.[198] The functional approach to the meaning of undertaking 'makes it possible to determine the categories of actors to which the competition rules apply'.[199] This sphere allocation function is made clear in

[191] See Section III above. [192] See Section III.A above.
[193] See Section III.C above. [194] Fuller (1969: 49–51).
[195] The lack of legal certainty is objected to in Lasok (2004).
[196] Kennedy (1982: 1349). [197] See Section IV.A.2 above.
[198] Kennedy (1982: 1349).
[199] Case C-67/96 *Albany International BV* [1999] ECR I-5751, AG Opinion para 206.

Wouters when the Court draws attention to the fact that the question is whether an entity 'is to be treated as an association of *undertakings* or, on the contrary, as a public authority' and in the jurisprudence treating regulation as public sphere activity under the free-movement rules.[200] Recognition of the sphere allocation function is important because it makes plain that the question is not whether an entity is or is not subject to Community law, but which are the appropriate provisions of Community law to apply. It enables a realization that though activities are *not* subject to the Community competition rules there is the very real possibility that they *are* subject to some other provision of Community law.[201]

Finally, acknowledging the need for a functional approach still leaves the question of what functions should be subject to Community competition law. Whilst clear that economic activity is subject to Community competition law, an implication of the functional approach taken in relation to undertakings is the possibility that the competition rules can apply to activity other than economic in relation to associations.[202] Attention is drawn to a question of whether the correct functions are subject to the Community competition law rules. The issue is complex and this chapter deals with it only briefly. However, Article 81 EC should not capture the activity of associations of undertakings that would lie beyond the scope of Article 81 EC if engaged in by undertakings. This narrow scope of the competition rules, as determined by the meaning of undertaking and economic activity, ensures that Article 81 EC is only applied to those engaged in activity in which not only does efficiency matter, but efficiency matters most. As most recently and most forcefully noted by Advocate General Maduro, 'In seeking to determine whether an activity carried on by the State or a State entity is of an economic nature, *the Court is entering dangerous territory*, since it must find a balance between the need to protect undistorted competition on the common market and respect for the powers of the Member States.'[203] The danger is apparent; too broad a conception of economic activity 'would represent an unlimited extension of the scope of competition law.'[204]

[200] Case C-309/99 *Wouters* [2002] ECR I-1577, para 56, emphasis added. Also AG opinion para 70, Joined Cases C-180/98 to C-184/98 *Pavel Pavlov* [2000] ECR I-6451, para 87–88, and Cruz (2002: 148–149, 154) and text accompanying notes 53–57 above.

[201] See Section IV.A.2 above and Drijber (2005: 530).

[202] ibid, para 27.

[203] Case C-205/03 P *FENIN*, AG Opinion para 26, emphasis added.

[204] Case C-205/03 P *FENIN*, AG Opinion para 27.

4

Collusion: Agreement and concerted practice

I. Introduction

The concept of collusion plays a central role in drawing the boundary of Article 81 EC. Article 81 EC only applies to *agreements* between undertakings, *decisions* of associations of undertakings, and *concerted practices*.[1] Agreements, decisions, and concerted practices are forms of collusion. Collusive behaviour is a precondition for the application of Article 81 EC; in *Bayer*, the CFI held that 'the prohibition ... concerns exclusively conduct that is coordinated *bilaterally* or *multilaterally*'.[2] Emphasis on collusion also exists in the equivalent US antitrust provision, so that 'condemnation has depended and continues to depend on finding two or more parties who may be said to have "agreed" to do what was done'.[3] The forms of collusion—agreement, decision, and concerted practice—are left undefined in the Treaty. The Court has attempted to provide definition through the cases, but a satisfactory conception eludes. Instead, the result is a considerable degree of confusion. Lidgard is concerned that the meaning of the collusive terms has 'not been addressed in a clear and systematic way'.[4] Wessely is of the view that '[t]he notion of collusion lacks any clear contours'.[5] Black writes that the terms have 'come to lack conceptual unity', are applied or described 'in various, obscure, confused and arguably contradictory ways', and that 'both the European Court of Justice and the domestic courts have

[1] It is clear that Article 81 EC also covers agreements between associations of undertakings: Joined Cases 209 to 215 and 218/78 *Heintz Van Landewyck SARL and Others v Commission (FEDETAB)* [1980] ECR 3125, para 48–56.

[2] Case T-41/96 *Bayer AG v Commission* [2000] ECR II-3383, para 64.

[3] Turner (1962: 655–656). Also Kovacic (1993: 5–15).

[4] Lidgard (1997: 352).

[5] Wessely (2001: 759).

fumbled with the concept of [concerted practice]'.[6] Jakobsen and Broberg consider the law so 'unreasonably unclear' that it is 'virtually impossible to draw the line between, on the one hand, unlawful (reciprocal) agreements and, on the other, lawful unilateral acts'.[7] The position is only marginally better in the United States, so that after seventy-four pages of analysis, Kovacic writes that '[d]espite a half-century of effort, antitrust law and policy have achieved only modest success in devising a satisfactory definition for the concept of concerted action and creating a suitable methodology for establishing the existence of an agreement'.[8] All that seems to be clear is that unilateral action is 'not an "agreement" by any stretch of the imagination'.[9]

Advocate General Darmon expressed the view that 'Courts must not engage in academic theorising. However, one is not succumbing to the temptation to theorise if one seeks to clarify certain concepts where the requirements of legal certainty are at stake.'[10] The aim of this chapter is to identify and develop the conceptions of agreement and concerted practice so as to arrive at a more certain position as to the boundaries of Article 81 EC than hitherto exists. As it is more certain than the concept of concerted practice, Section II involves examination of the concept of agreement, finding that common intention is required. Separate from the meaning of agreement is how agreement is shown to exist. The framework of offer and acceptance is used to clarify how common intention is evidenced. Section III then provides an examination of concerted practice: two conceptions are considered. The first treats concerted practice as an alternative method of demonstrating common intention. The second treats concerted practices as methods of reducing uncertainty about future conduct. This conception of concerted practice is evidenced by communication.

It is argued that like the concept of undertaking examined in the previous chapter, the collusive concepts can be seen to serve a jurisdictional rather than substantive function.[11] The existence of collusion is separate from the question of whether the collusion causes harm to competition. Falling within the scope of Article 81 EC does not mean that conduct is prohibited by Article 81 EC, but simply indicates that Article 81 EC contains the

[6] Black (2003d: 504, 505) and Black (1992: 200).
[7] Jakobsen and Broberg (2002: 139).
[8] Kovacic (1993: 80).
[9] Turner (1962: 658). Also Black (2003c) and Hylton (2003: 140).
[10] Joined Cases C-89/85, C-104/85, C-114/85, C-116/85, C-117/85 and C-125/85 to C-129/85 *A. Ahlström Osakeyhtiö v Commission* [1993] ECR I-1307, AG Opinion para 166.
[11] Waelbroeck and Frignani (1999: 117–154) classify the collusive concepts as rules of substance.

appropriate rules of substantive assessment. This, and the accompanying clarification that there is nothing pejorative in the term collusion, makes the broad reach of the conception of collusion identified and advocated tolerable. Section IV considers the limit of the jurisdiction claimed by the collusive terms by examining the extent to which outcomes matching those that would be reached if there were collusion can be considered collusive. It is recognized that reaching the *as if* colluded outcome without colluding is *not* prohibited as collusion and so falls outside the boundary of Article 81 EC. However, the ability to reach the *as if* colluded outcome without colluding is subject to scrutiny. Conclusions are drawn in Section V.

II. Agreement

It is recognized that 'in the early cases no definitional problem of what "agreement" meant even arose. The parties to formal cartel agreements dutifully notified their arrangements to the Commission.'[12] As undertakings learnt that certain types of agreement were prohibited, they sought to conceal their existence.[13] The challenge is then to find a conception of agreement that is capable of being proven when the undertakings do not openly acknowledge, and actively seek to conceal, its existence.[14] Section II.A explores the meaning of agreement within Article 81 EC.[15] Section II.B considers how the existence of an agreement can be evidenced or proven. Finally, Section II.C considers how such evidence is obtained.

A. *The meaning of agreement within Article 81 EC*

The meaning of agreement in Article 81 EC can be seen to share the same meaning as ascribed to agreement in the law of contract.[16] As described

[12] Joshua and Jordan (2004: 657). Similar US experience is described in Gellhorn *et al.* (2004: 266–268) and Sullivan and Grimes (2000: 177).

[13] Joshua and Jordan (2004: 657–658), Kovacic (1993: 17–18) and Van den Bergh and Camesasca (2001: 194).

[14] Joshua and Jordan (2004: 654–655), Whish (1985: 23), and Whish (1993: 43).

[15] Compare with Black (2005: Ch 4).

[16] Contract generally involves legally enforceable agreements: Beatson (2002: 1–2) and Treitel (2004: 1–3). However, contract and agreement are not synonymous. There can be contract without agreement and agreement without contract: Bayles (1987: 147–148, 151–152), Beale (2004: para 1-003–1-005), Black (2004a), Buckley (2005: 22–26), Campbell (2001), Fried (1981: 3–5, 10–25), Fuller and Perdue Jr (1936a), Fuller and Perdue Jr (1936b), Posner (2003: 96,

by Beatson, 'contract... involves at least two parties... and an outward expression of *common intention*'.[17] Fried writes that 'the focus of the inquiry is on the *will of the parties*'.[18] As in contract, Article 81 EC takes common intention to be central to the meaning of agreement.[19] In a section headed 'meeting of the minds', Waelbroeck and Frignani write that 'Article [81 EC] prohibits anticompetitive behavior only if it is the result of the concurring will of two or more undertakings'.[20] Goyder is of the view that for the purpose of Article 81 EC, agreement is reached 'when there can objectively be said to be sufficient consensus'.[21] The CFI judgment in *Bayer* confirms this conception of agreement, since:

> in order for there to be an agreement within the meaning of Article [81(1)] of the Treaty it is *sufficient* that the undertakings in question should have expressed their *joint intention* to conduct themselves on the market in a specific way.[22]

Not only is common intention *sufficient* for agreement within the meaning of Article 81 EC, common intention is *necessary* for there to be an agreement within the meaning of Article 81 EC, since:

> [t]he proof of an agreement... within the meaning of Article [81(1)] of the Treaty *must be founded* upon the direct or indirect finding of the existence of the *subjective element* that characterises the very concept of an agreement, that is to say a *concurrence of wills* between economic operators.[23]

135–136), Shavell (2004: 299–304, 335–337, 310, 313–314, 317), Smith (2004: 44, 60–68, 78–97), Treitel (2004: 2–5), Tullock (1971: 51–55, 320–322), and Williamson (1979a: 238–240). Honig *et al.* (1963: 9–10) caution against equating the meaning of agreement in Article 81 EC *with* contract. What is intended here is to equate the meaning of agreement in Article 81 EC with the meaning of agreement *in* contract.

[17] Beatson (2002: 27), emphasis added.

[18] Fried (1981: 4), emphasis added. Also Beale (2004: para 2-002–2-003, 2-0025–2-042), Black (2003d: 506), Black (2004b: 80–83), and Martin (2001: 49).

[19] Though Ibbetson (1999: 245–261) and Buckley (2005: 27–34) critique the idea that *contract* flows from common intention, this does not detract from the idea that *agreement* flows from common intention.

[20] Waelbroeck and Frignani (1999: 121). Also Case C-551/03 P *General Motors*, AG Opinion para 36.

[21] Goyder (2003: 66).

[22] Case T-41/96 *Bayer AG v Commission* [2000] ECR II-3383, para 67, emphasis added, citing Case 41/69 *ACF Chemiefarma v Commission* [1970] ECR 661, para 112, Joined Cases 209/78 to 215/78 and 218/78 *Van Landewyck and Others v Commission* [1980] ECR 3125, para 86, and Case T-7/89 *Hercules Chemicals v Commission* [1991] ECR II-1711, para 256.

[23] Case T-41/96 *Bayer AG v Commission* [2000] ECR II-3383, para 173, emphasis added, also 69, Case T-1/89 *Rhône-Poulenc SA v Commission* [1991] ECR II-867, para 120, Case T-2/89 *Petrofina v Commission* [1991] ECR II 1087, para 211, Case T-7/89 *Hercules Chemicals NV SA v Commission* [1991] ECR II-1711, para 256, and Faull and Nikpay (1999: para 2.27).

B. *Proof of agreement*

Common intention is central to the meaning of agreement in both Article 81 EC and contract. Common intention in Article 81 EC can be shown to exist in the same way that common intention is shown to exist in contract. In contract, common intention is normally evidenced by finding a definite offer by one party, and an equally definite acceptance of that offer by another party.[24] This can be seen as requiring communication (offer) and commitment (acceptance). The framework of offer (communication) and acceptance (commitment) can be used to clarify the jurisprudence of the Court on how common intention is proven for the purpose of Article 81 EC.[25] A number of things are made plain within this framework.

The existence of an offer, and of the acceptance of that offer, may be shown by written evidence. In vertical relations there will often be a formal written contract evidencing common intention; rarely will this be so for clearly prohibited practices.[26] The existence of an offer and of its acceptance is often established from exchanges of correspondence, minutes of meetings, and other such documents.[27] This can be seen from recent UK investigations into collusive tendering. Collusive tendering is a method of removing customer choice in relation to supplier and price. Competitors decide the winner of the bid and the price in advance. The remaining undertakings submit token bids that are too high; the purchaser is given the impression that the cheapest supplier is offering a competitive price.[28] In *Collusive Tendering in Relation to Contracts for Flat Roofing Services in the*

[24] Beatson (2002: 27–28), Martin (2001: 49), and Neven *et al.* (1998: 57).

[25] Compare with Black (2003d: particularly 505) and Case 1021/1/1/03 and 1022/1/1/03 *JJB Sports PLC v Office of Fair Trading; Allsports Limited v Office of Fair Trading* [2004] CAT 17, para 868. In US antitrust commentators look for an exchange of assurances: Kovacic (1993: 18).

[26] Joshua and Jordan (2004: 654–655) and Whish (1993: 43).

[27] Case C-199/92 P *Hüls AG v Commission* ECR [1999] I-4287, para 141–155, Case T-1/89 *Rhône-Poulenc SA v Commission* [1991] ECR II-867, AG Opinion 955–957, Gavil *et al.* (2002: 244–245), van Gerven and Navarro Varona (1994: 605–607, though in the context of concerted practice), Guerrin and Kyriazis (1993: 790, 802–804) and Wessely (2001: 740). In Joined Cases 100–103/80 *Musique Diffusion Française SA v Commission* [1983] ECR 1825, AG Opinion 1915, it was sufficient that Shriro possessed an unsigned standard form contract from Pioneer NV.

[28] Case CP/0001–02 *Collusive Tendering in Relation to Contracts for Flat Roofing Services in the West Midlands* [2004] 16 March, para 18–21. It was standard practice for roofing contractors to request a price from competitors to enable them to put in a tender for work they had no interest in winning so as to remain on the tender list: Case 1033/1/1/04 *Richard W Price (Roofing Contractors) Limited v Office of Fair Trading* [2005] CAT 5, para 9, 37, 54 and Case 1032/1/1/04 *Apex Asphalt and Paving Co Limited v Office of Fair Trading* [2005] CAT 4, para 248–253.

West Midlands, written offers to submit bids of a specified value in response to an invitation to tender were made and discovered.[29]

The existence of an offer, and of the acceptance of that offer, may be established by parol evidence.[30] It is enough that words are said, so long as they can be evidenced.[31] Both written and parol evidence of offer and acceptance may come from a third party.[32] Acceptance need not be established by the same type of evidence as the offer. The existence of an offer, and of the acceptance of that offer, may also be inferred from other forms of conduct. This requires a more detailed explanation and is considered in Section II.B.1 below. In the event of inconsistent evidence, questions may arise as to whether there is a hierarchy of evidence between parol, written, and other types of conduct evidencing common intention.[33]

Reasons why undertakings make offers or accept them are separate from the question of whether there has been offer or acceptance.[34] Threats may be used to pressurize undertakings to engage in certain conduct.[35] Coercion *could* be distinguished from offer in that nasty consequences are promised for failure to accede.[36] However, Article 81 EC does not make the distinction.[37] It is also clear that coercion does not vitiate acceptance; it

[29] Case CP/0001-02 *Collusive Tendering in Relation to Contracts for Flat Roofing Services in the West Midlands* [2004] 16 March, para 36–40, 157–161, 176–177, 208–209.

[30] Case T-1/89 *Rhône-Poulenc SA v Commission* [1991] ECR II-867, AG Opinion 954–955, and Guerrin and Kyriazis (1993: 816–818).

[31] Case 28/77 *Tepea BV v Commission* [1978] ECR 1391, para 17, 40–41, Case 1021/1/1/03 and 1022/1/1/03 *JJB Sports PLC v Office of Fair Trading; Allsports Limited v Office of Fair Trading* [2004] CAT 17, para 510–511, 558–559, and Joined Cases 100-103/80 *Musique Diffusion Française SA v Commission* [1983] ECR 1825, AG Opinion 1915.

[32] Joined Cases 40 to 48, 50, 54 to 56, 111, 113 and 114–73 *Coöperatieve Vereniging "Suiker Unie" Ua v Commission* [1975] ECR 1663, para 132–140, 159–166, 237–238, 271–274, AG Opinion 2081 (though the written hearsay evidence is actually used to establish a concerted practice), Case CP/0871/01 *Price-Fixing of Replica Football Kit* [2003] 1 August, para 344, 349–354, 369, 371, 374–380, Case T-1/89 *Rhône-Poulenc SA v Commission* [1991] ECR II-867, AG Opinion 954–960, and Case C-199/92 P *Hüls AG v Commission* [1999] ECR I-4287, para 143, 147.

[33] Case CP/0871/01 *Price-Fixing of Replica Football Kit* [2003] 1 August, para 357 and Dashwood *et al.* (1976: 498–499). Gellhorn *et al.* (2004: 269) write that 'Documents and testimony typically stand atop the hierarchy of proof because they tend to give the courts greater confidence that the defendants acted jointly.'

[34] Compare with Smith (2004: 315–323).

[35] Posner (2003: 137).

[36] Smith (2004: 318–319).

[37] Case 107/82 *AEG-Telefunken v Commission* [1983] ECR 3151, AG Opinion 3230, Case T-342/99 *Airtours v Commission* [2002] ECR II-2585, para 183–207, Rivas and Branton (2003: 1235), Turner (1962: 684–695), Martin (2001: 50), and Posner (2003: 288–289). The question of coercion is relevant at the sanctioning stage: Whish (2003: 93), Turner (1962: 700), and Jakobsen and Broberg (2002: 134–135).

does not matter why an offer is accepted, just that it is accepted—even if unwillingly.[38]

Finally, there is no need for an agreement within the meaning of Article 81 EC to meet all the requirements of contract or for the agreement to be legally binding.[39] This is established in *Sandoz*, where goods supplied were accompanied by an invoice containing standard terms, which included the words 'export prohibited'.[40] The Commission considered this to be part of the agreement between Sandoz and its customers.[41] Sandoz argued that as Article 81(2) EC renders anti-competitive agreements unenforceable, the terms were not binding, hence not part of the agreement for the purpose of Article 81 EC. The Court confirmed that it is unnecessary for an agreement 'to constitute a valid and binding contract under national law'.[42] All that is required is evidence of a faithful expression of intention.[43]

1. *Inference from conduct*

'When the event we wish to study is clandestine, we cannot rely upon direct observation.'[44] Thus, it is possible to infer both the existence of an offer and

[38] Case T-9/89 *Hüls v European Commission* [1992] ECR II-499, para 128, Case 16/61 *Modena v High Authority* [1962] ECR 289, 303, Joined Cases 100-103/80 *Musique Diffusion Française SA v Commission* [1983] ECR 1825, paras 88–90, 100, Case CP/0871/01 *Price-Fixing of Replica Football Kit* [2003] 1 August, para 406, Jakobsen and Broberg (2002: 134), Waelbroeck and Frignani (1999: 123–124), and Turner (1962: 685–691, 694, 700–701). Compare with Smith (2004: 324–326) in relation to contract. For evidence and acknowledgement of coercion: Case CP/0871/01 *Price-Fixing of Replica Football Kit* [2003] 1 August, para 112, 120, 125, 158, 166, 168, 172, 176, 183, 193, 206–209, 260, 388, 391, 397, 400, 480–482, 484, 497–500 and Case 1021/1/1/03 and 1022/1/1/03 *JJB Sports PLC v Office of Fair Trading; Allsports Limited v Office of Fair Trading* [2004] CAT 17, para 85, 351–354, 358–360, 365, 394–503. Some parties were able to resist the coercive pressure: Case CP/0871/01 *Price-Fixing of Replica Football Kit* [2003] 1 August, para 169, 415. In closely prescribed circumstances coercive pressure from the state does remove the element of voluntariness necessary for acceptance: Case T-387/94 *Asia Motor France SA v Commission* [1996] ECR II-961, para 43–72, particularly para 61, 65, Case C-198/01 *Consorzio Industrie Fiammiferi (CIF) v Autorita Garante della Concorrenza e del Mercato* [2003] ECR I-8055, Temple Lang (2004), and Waelbroeck and Frignani (1999: 142–146).

[39] Joined Cases 209 to 215 and 218/78 *Heintz Van Landewyck SARL and Others v Commission (FEDETAB)* [1980] ECR 3125, para 85–86, Whish (2003: 92), Jones and Sufrin (2004: 130–131) and note 16 above. Compare Kovacic (1993: 16–17). This is not an uncontested view: Section III.A.1. (a) below.

[40] Sandoz [1987] OJ L 222/28 [1989] 4 CMLR 628, para 20.

[41] Sandoz [1987] OJ L 222/28 [1989] 4 CMLR 628, para 10–15, 25–28.

[42] C-277/87 *Sandoz Prodotti Farmaceutici SpA v Commission* [1990] ECR I-45, para 2 of the summary and Waelbroeck and Frignani (1999: 122–123). The full judgment is not reported. Compare with Volkswagen [2001] OJ L 262/14, overturned in Case T-208/01 *Volkswagen v Commission* [2003] ECR II-5141, which itself is an appeal as Case C-74/04 P.

[43] Joined Cases 209 to 215 and 218/78 *Heintz Van Landewyck SARL and Others v Commission (FEDETAB)* [1980] ECR 3125, para 85–86.

[44] Stigler (1983a: 268).

of its acceptance from conduct in the market place.[45] As the considerations that arise depend on whether an offer or its acceptance is being inferred, they are treated separately:[46]

(a) Inference of offer from conduct The circumstances in which it is appropriate to infer an offer from conduct are clarified in *Bayer*.[47] Bayer AG, a chemical and pharmaceuticals company, produced Adalat, the trade-name for a calcium antagonist used to treat cardio-vascular diseases, which it then sold to wholesalers in France, Spain, and the UK.[48] Between 1989 and 1993, the price of Adalat (fixed by the Spanish and French health authorities) was between 24 per cent and 55 per cent lower in France and Spain than the price prevailing in the UK.[49] This created an incentive to purchase in France or Spain for resale in the UK.[50] In response to this trend, Bayer monitored the market to identify the source of the imports; Bayer then reduced supplies to wholesalers in Spain and France—sometimes supplies were reduced by 90 per cent.[51] Since French and Spanish law required wholesalers to give priority of supply to the national market, the reduction in supplies reduced exports.[52]

There was neither written nor parol evidence that the offer to supply goods was conditional on them not being exported.[53] The Commission attempted to infer the existence of the offer from conduct.[54] Bayer argued,

[45] Case T-1/89 *Rhône-Poulenc SA v Commission* [1991] ECR II-867, AG Opinion 957–960, Joined Cases 100-103/80 *Musique Diffusion Française SA v Commission* [1983] ECR 1825, AG Opinion 1930 speaking of concerted practice, Black (2003d: 505), Broder (2005: para 3–007), Gavil *et al.* (2002: 245–267), Gellhorn *et al.* (2004: 268–283), Hylton (2003: 132–140), Joshua (1987: 318–320), Posner (1979a: 179–184), and Van den Bergh and Camesasca (2001: 201–202).

[46] Common intention inferred from conduct is characterized by some as a concerted practice, rather than as an agreement and is further discussed in Section III.A.1.(b) below.

[47] Noted by Lidgard (1997). Also Case C-551/03 P *General Motors BV and Opel Nederland BV*, AG Opinion para 24–27.

[48] Adalat [1996] OJ L 201/1, para 3, 8–12, 37 and Case T-41/96 *Bayer AG v Commission* [2000] ECR II-3383, para 22, 27.

[49] Adalat [1996] OJ L 201/1, para 29–31 and Case T-41/96 *Bayer AG v Commission* [2000] ECR II-3383, para 31–32.

[50] Adalat [1996] OJ L 201/1, para 32–35 and Case T-41/96 *Bayer AG v Commission* [2000] ECR II-3383, para 30–33.

[51] Adalat [1996] OJ L 201/1, para 32–35, 53–87, 109–111, 138–141, 147 and Case T-41/96 *Bayer AG v Commission* [2000] ECR II-3383, para 34–37.

[52] Adalat [1996] OJ L 201/1, para 40, 44, 104–108, 115–121, 132–137 and Case T-41/96 *Bayer AG v Commission* [2000] ECR II-3383, para 60.

[53] Adalat [1996] OJ L 201/1, para 50, 52 and Case T-41/96 *Bayer AG v Commission* [2000] ECR II-3383, para 73, 123.

[54] Adalat [1996] OJ L 201/1, para 89, 93, 96, 156–188 (particularly para 176–188) and Case T-41/96 *Bayer AG v Commission* [2000] ECR II-3383, para 48–51, 75, 158–171.

and the CFI accepted, that because the strategy did not require participation by, or assistance from, other parties, it was wrong to infer that others were being invited to participate in the strategy.[55]

(i) Unilateral conduct *Bayer* confirms that the unilateral action is 'not an "agreement" by any stretch of the imagination'.[56] Conduct is considered unilateral when its aims can be achieved 'without the express or implied participation of another undertaking'.[57] It is not possible to infer an offer from conduct when the aims of the practice can be achieved without the assistance of others.[58]

(ii) Apparently unilateral conduct Some conduct appears unilateral but its aims *cannot* be achieved without the assistance of others. Such conduct is termed apparently, as opposed to genuine, unilateral conduct.[59] It is possible to infer an offer from apparently unilateral conduct. In *AEG*, AEG could not achieve its aims of high retail pricing unilaterally, but needed the cooperation of retailers. Cooperation was obtained by coercive pressure in the form of 'a threat to break off business relations'.[60] Thus, the Advocate General found that, after heated discussion over discounts offered by a retailer and the threat of 'a "considerable worsening of relations"... "the likelihood of a renewed aggressive advertising campaign" had been reduced'.[61] The Court concurred in finding that action was not unilateral:

> [o]n the contrary, it forms part of the contractual relations between the undertaking and resellers.... approval is based on the acceptance, tacit or express, by the contracting parties of the policy pursued by AEG which requires *inter alia* the

[55] Adalat [1996] OJ L 201/1, para 86–108 and Case T-41/96 *Bayer AG v Commission* [2000] ECR II-3383, para 38, 76, 81–110. The Commission did not accept that the strategy could be successful without cooperation from the wholesalers: Case T-41/96 *Bayer AG v Commission* [2000] ECR II-3383, para 52, 112.

[56] Turner (1962: 658).

[57] Case T-41/96 *Bayer AG v Commission* [2000] ECR II-3383, para 71. Also Black (2003c: 103) and Hylton (2003: 135–136, 140–143).

[58] Case T-41/96 *Bayer AG v Commission* [2000] ECR II-3383, para 119–122. At para 71 and 111 the CFI adds confusion by suggesting that unilateral conduct becomes merely apparently unilateral conduct when and if it receives at least tacit acquiescence. This is erroneous; it is impossible to accept an offer that has not been made. Compare with Case 107/82 *AEG-Telefunken v Commission* [1983] ECR 3151, AG Opinion 3257 and Lidgard (1997: 359).

[59] Case T-41/96 *Bayer AG v Commission* [2000] ECR II-3383, para 70 and Lidgard (1997: 356–358).

[60] Case 107/82 *AEG-Telefunken v Commission* [1983] ECR 3151, AG Opinion 3254.

[61] Case 107/82 *AEG-Telefunken v Commission* [1983] ECR 3151, AG Opinion 3261, para 107–134 of the judgment give the courts finding of improper pressure.

exclusion from the network of all distributors who are . . . not prepared to adhere to that policy.[62]

In addition to coercive pressure, apparently unilateral conduct may involve 'cheap talk'. This entails announcing a strategy (making an offer) long before implementation is required. The announced position becomes a focal point.[63] If others do not accept the strategy, the strategy need not be implemented.[64] The Court has treated the announcement of focal points as offers.[65] For example, in *Dyestuffs*, undertakings would announce price increases long before they were due to take effect; at a meeting held on 18 *August* 1967 one undertaking announced 'that it would increase prices of dyestuffs by 8 per cent on 16 *October* 1967'.[66] According to Advocate General Myras, 'everything happens as if the price leader intended to give the other producers a period of reflection'.[67] The conduct is not unilateral because the price increase 'can only be put into effect if it is accepted by the other producers'.[68] If others do not accept the offer, the party announcing a price increase need not put it into effect.[69]

'Cheap talk' was alleged by the Commission in *Wood pulp*, it being argued that:

[b]y announcing their prices well before they came into effect at the beginning of a quarter, the producers gave sufficient time to their competitors to bring their own announced prices into line.[70]

However, the Court was unable to infer an offer that was capable of being accepted from the price announcements. One reason was that orders could

[62] Case 107/82 *AEG-Telefunken v Commission* [1983] ECR 3151, para 38, also AG Opinion 3230, 3258. At AG Opinion 3250, the Commission considered that an agreement can be established 'by unilateral campaigns by the applicant ranging from . . . recommendations and intensive . . . discussions to instructions, accompanied by hidden pressure'.

[63] Scherer and Ross (1990: 265–268) and Van den Bergh and Camesasca (2001: 177).

[64] Farrell and Rabin (1996) and Motta (2004: 152–156).

[65] In Case 48/69 *Imperial Chemical Industries v Commission* [1972] ECR 619, 678, AG Myras felt it unnecessary 'to describe this operation in relation to the law of contract'.

[66] Re Cartel in Aniline Dyes [1969] CMLR D23, para 9, 91–99, AG Opinion 679, emphasis added. This 'fact' is subject to comment in Mann (1973: 41).

[67] Case 48/69 *Imperial Chemical Industries v Commission* [1972] ECR 619, 678.

[68] Case 48/69 *Imperial Chemical Industries v Commission* [1972] ECR 619, AG Opinion 678. Acceptance of this offer would then complete the elements of agreement.

[69] Case 48/69 *Imperial Chemical Industries v Commission* [1972] ECR 619, AG Opinion 678. Compare with Rostow (1947: 581–582).

[70] Joined Cases C-89/85, C-104/85, C-114/85, C-116/85, C-117/85 and C-125/85 to C-129/85 *A. Ahlström Osakeyhtiö v Commission* [1993] ECR I-1307, 1326, 1364–1368, judgment para 60.

be placed as soon as prices were announced, so there was no time to retract an offer if it was not accepted by competitors.[71]

(b) Inference of acceptance from conduct When an offer is made, be it parol, written, or inferred from conduct, compliance with the terms of the offer is treated as acceptance.[72] This is clear from the OFT *Collusive Tendering in Relation to Contracts for Flat Roofing Services in the West Midlands* decision. Written offers to submit bids of a specified value in response to an invitation to tender were made.[73] These offers were accepted by those who submitted bids of the specified value.[74] There is always the question of whether the conduct is referable to the offer, and thus constitutive of acceptance.[75] However, from the figures submitted, there was no reasonable explanation for the conduct other than that they were responses to the offers.[76]

In *Bayer*, the Commission felt that an offer to supply goods subject to an export prohibition was made, and that those supplied knew that these were the terms of business; by continuing relations, there was acceptance by conduct.[77] Acceptance of an offer can and has been inferred from continuing relations.[78] However, contrary to the Commission's findings, there was simply no offer to accept.[79]

(i) Adherence to the agreement For the purpose of Article 81 EC, it is normally irrelevant whether parties adhere to an agreement or intend to adhere to an agreement.[80] This is clearly correct when the evidence of

[71] Joined Cases C-89/85, C-104/85, C-114/85, C-116/85, C-117/85 and C-125/85 to C-129/85 *A. Ahlström Osakeyhtiö v Commission* [1993] ECR I-1307, 1365.

[72] Jakobsen and Broberg (2002: 128–131) and Case CP/0871/01 *Price-Fixing of Replica Football Kit* [2003] 1 August, para 306. Faull and Nikpay (1999: para 2.49) suggest this is only sufficient to establish a concerted practice. Compare with Hylton (2003: 133–143) and Posner (2003: 102).

[73] See references cited at note 29 above.

[74] Case CP/0001–02 *Collusive Tendering in Relation to Contracts for Flat Roofing Services in the West Midlands* [2004] 16 March, para 41–42, 161–163, 178–179, 210–211.

[75] Smith (2004: 183–187).

[76] Case CP/0001-02 *Collusive Tendering in Relation to Contracts for Flat Roofing Services in the West Midlands* [2004] 16 March, para 41–42, 164.

[77] Adalat [1996] OJ L 201/1, para 89, 93, 96, 156–188 (particularly para 176–188) and Case T-41/96 *Bayer AG v Commission* [2000] ECR II-3383, para 48–51, 75, 158–171.

[78] Whish (2003: 103–105).

[79] Adalat [1996] OJ L 201/1, para 171 and references cited at note 58 above.

[80] Case 107/82 *AEG-Telefunken v Commission* [1983] ECR 3151, AG Opinion 3251, 3259, Case T-305/94 *Limburgse Vinyl Maatschappij v Commission* [1999] ECR II-931, para 773, Case T-141/89 *Tréfileurope v Commission* [1995] ECR II-791, para 60, 85, Case C-246/86 *Belasco v Commission* [1989] ECR 2117, para 10–16, Case CP/0001-02 *Collusive Tendering in Relation to*

acceptance is written or parol.[81] So, in *Price-Fixing of Replica Football Kit*, it is of no moment that Sports Soccer's conduct continually deviates from that which is agreed; the existence of agreement is established by written and parol evidence.[82] However, when inferring acceptance from conduct it seems logical that the conduct *must be consistent* with the offer. Therefore, in *Bayer* it is significant that the wholesalers did not act consistently with the alleged offer; even if an offer did exist, it is not possible to infer acceptance from the conduct of the wholesalers.[83]

C. *Obtaining evidence of agreement*

Knowing what agreement is, and what is required to show agreement, still leaves the task of obtaining evidence of agreement.[84] Some agreements are contained in contract; the existence becomes apparent when one of the parties seeks to enforce the contract through the courts.

Under Regulation 17/62, parties were required to admit the existence and terms of an agreement when asking the Commission to either certify that there is no basis to act under Article 81(1) EC or to confer the benefit of Article 81(3) EC on a particular agreement.[85] The replacement of Regulation 17/62 with Regulation 1/2003 removed this source of information.[86] However, it is doubtful whether notification was a useful

Contracts for Flat Roofing Services in the West Midlands [2004] 16 March, para 191, Case Comp/C.37.750/B2 Co Brasseries Kronenbourg—Brasseries Heineken, para 61, 64, and Jakobsen and Broberg (2002: 135).

[81] Subject to questions of hierarchy of evidence, discussed at note 33 above.

[82] Case CP/0871/01 *Price-Fixing of Replica Football Kit* [2003] 1 August, para 355–356, also inferred from conduct at para 346–348. Conduct deviates from that required by the agreement at para 193–194, 200, 221, 233, 237–240, 359–361, 381, 390, 396, and Case 1021/1/1/03 and 1022/1/1/03 *JJB Sports PLC v Office of Fair Trading; Allsports Limited v Office of Fair Trading* [2004] CAT 17, para 73 and 82.

[83] *Adalat* [1996] OJ L 201/1, para 96–103, Case T-41/96 *Bayer AG v Commission* [2000] ECR II-3383, para 38, 125–157, and Jakobsen and Broberg (2002: 131–132).

[84] Guerrin and Kyriazis (1993: 775–777, 791–792).

[85] Green Paper on Vertical Restraints in EC Competition Policy Com (96) 721 Final [1997] 4 CMLR 519, para 188–192, Regulation 17. First Regulation Implementing Articles 81 and 82 of the Treaty [1962] OJ Special Edition 204/62, Article 4(1), European Commission (1997), Ortiz Blanco (1996: 64–65, 68, 69–70), Van den Bergh and Camesasca (2001: 200–202), and Wesseling (2004: 1141, 1147–1149).

[86] Article 10 (clarified by Recital 14) of Regulation 1/2003 allows the Commission to consider the application of Article 81 EC in relation to new types of practices. This source of inward information is thus not entirely exhausted: Whish (2003: 257) and Wils (2005: §1.1.3 n 22).

source of information, since practices with the most serious anti-competitive consequences were so rarely notified.[87]

The Commission has power to actively gather information.[88] Article 18 of Regulation 1/2003 enables the Commission to demand information necessary for an assessment of market practices from governments, national competition authorities, undertakings, and associations.[89] Articles 20–22 of Regulation 1/2003 enable investigations to obtain information necessary to reach an assessment of practices in relation to Article 81.[90] There are limitations to what may be revealed by investigation, since '[t]he smoothly functioning cartel is less likely to generate evidence of actual agreement'.[91] The challenge remains to 'penetrate [the] cloak of secrecy.'[92]

Investigative powers used to locate evidence of the existence and content of agreements are enhanced by a power to exploit 'a natural nervousness within the ranks of cartels by holding out a tempting offer of immunity or leniency for the first, but only the first, to provide evidence'.[93] Leniency is an aid to existing investigative tools: one escapes the sanction, in exchange for assisting in obtaining evidence against a wider group.[94] This relies on 'a genuinely good offer: be the first to blow the whistle and there will be complete immunity from a big penalty—a big fine and prison sentences for sure when (and not if) the cartel is found out'.[95] Article 83(1)(a) EC empowers the Council to 'ensure compliance with the prohibitions laid down in Article 81(1) and in Article 82 by making provision for fines and

[87] Faull and Nikpay (1999: para 6.09), Guerrin and Kyriazis (1993: 783), Van den Bergh and Camesasca (2001: 194–195), Wils (2005: §1.1.4.2.2), and references cited at note 13 above. On the failings of notification more generally: Green Paper on Vertical Restraints in EC Competition Policy Com (96) 721 Final [1997] 4 CMLR 519, para 185, Goyder (1998: 46–51), Hawk (1995: 984) and Wils (2005: §1.1.4).

[88] Jephcott and Lübbig (2003: para 3.4–3.62, 3.100–3.169) and Harding and Joshua (2003: 164–169).

[89] This is a similar power to that previously available under Article 11 of Regulation 17/62: Wils (2005: §5.1.1).

[90] This is similar to the power that existed under Article 14 of Regulation 17/62, under which an investigation could result from an application for clearance or notification for exemption; complaint; parliamentary question; contact with other antitrust authorities; reports in newspapers and trade journals: Kerse (1998: para 3.01), European Commission (1997: para 2.3), Ortiz Blanco (1996: 97), and on the new powers Wils (2005: §5.1.1).

[91] Posner (2003: 289). [92] Jephcott and Lübbig (2003: para 5.4).

[93] Harding and Joshua (2003: 209, also 35–36, 213–228) and Jephcott and Lübbig (2003: para 5.1–5.67, 8.7–8.12).

[94] Office of Fair Trading (2005: 4), Office of Fair Trading (2004c: para 3.2), Rennie (2005: 165, 167, 173), Wils (2002a: 28–29, 37), and Wils (2005: §5.2.3).

[95] Harding and Joshua (2003: 215).

periodic penalty payments'. The Commission's power to impose fines was initially contained in Articles 15 and 16 of Regulation 17/62 and is now conferred by Articles 23 and 24 of Regulation 1/2003.[96] These provisions enable the Commission to impose a fine of up to 10 per cent of an undertaking's global turnover in the preceding business year: 'clearly this can be an enormous amount'.[97] The Commission will grant immunity from fines to the first undertaking to provide evidence sufficiently robust to prove an infringement without further investigation.[98] The approach has yielded success, with a large number of Commission cases now being based on evidence obtained from leniency applicants.[99]

III. Concerted practice

'Agreement' captures expressions of common intention that are shown to exist by evidence of communication (offer) and commitment (acceptance), be it written, parol, or inferred from conduct. It is acknowledged that the purpose of concerted practice is to capture forms of conduct not caught by 'agreement'.[100] In *Wood pulp*, after examining numerous cases and

[96] Regulation 17. First Regulation Implementing Articles 81 and 82 of the Treaty [1962] OJ Special Edition 204/62 and Council Regulation (EC) No 1/2003 [2003] OJ L 1/1. The power is to impose fines on undertakings and not individuals (unless the individual is also an undertaking): Kerse and Khan (2005: para 7–002), Smith (2001: para 9.03), and Wils (2000: 99–100). Zuleeg (1999: 448–450) is of the view that nothing in the EC Treaty prevents the introduction of individual sanctions. This is now given positive support by Case C-176/03 *Commission v Council* [2005] ECR, nyr.

[97] Whish (2003: 267).

[98] Commission Notice on Immunity from Fines and Reduction of Fines in Cartel Cases [2002] OJ C 45/3, para 8(b) and Kerse and Khan (2005: para 7-060). The ability to grant leniency seems to be an exercise of the discretion as to whether to impose a sanction and to determine its level: Joined Cases T-305/94, T-306/94, T-307/94, T-313/94, T-314/94, T-315/94, T-316/94, T-318/94, T-325/94, T-328/94, T-329/94 and T-335/94 *Limburgse Vinyl Maatschappij NV v Commission* [1999] ECR II-931, para 1173, Case T-347/94 *Mayr-Melnhof v Commission* [1998] ECR II-1751, para 246–285, Smith (2001: para 9.44, 9.49, 9.55), and Montag (1996: 432). The discretion is not unfettered: Article 229 EC, recital 33 and Article 31 of Council Regulation (EC) No 1/2003 [2003] OJ L 1/1, Article 17 of Regulation 17. First Regulation Implementing Articles 81 and 82 of the Treaty [1962] OJ Special Edition 204/62, Smith (2001: 214–215) and Roth QC (2001: para 12-128). Compare with the position in Australia: Rennie (2005: 171). The limits on discretion can also be compared with US courts' ability to review settlements entered into with enforcement agencies, although the consent order process is distinct from the leniency process: *United States v Western Elec. Co., Inc.*, 767 F. Supp. 308, 328–329 (D.D.C. 1991), Bodner Jr (1988), Kauper (1993), and Savin (1997).

[99] European Commission (2004: 181–213).

[100] Faull and Nikpay (1999: para 2.45), Rose (1993: para 2-034), Goyder (2003: 71–72), and Whish (2003: 99).

commentaries, Advocate General Darmon concluded that 'there is still, in my view, some uncertainty regarding the definition of concerted practices'.[101] This section seeks to both identify and develop a conception of concerted practice that captures conduct not caught within the meaning of agreement, but which shares a common element so that both can be seen as forms of collusion.[102] What constitutes evidence of a concerted practice is then considered, the meaning of concerted practice being distinct from questions of what evidences concerted practice.[103]

A. *The meaning of concerted practice within Article 81 EC*

Two conceptions of concerted practice are considered. The first requires common intention, but relies on different evidence than that used to show agreement. The second does not require common intention, but instead focuses on whether conduct reduces uncertainty as to the future conduct of others.

1. *Common intention*

One view is that concerted practice requires 'the deliberate and mutual adoption of a *common intention*'.[104] Advocate General Mayras expressed the view that the concept of concerted practice in Article 81 EC originates from the US antitrust conception of conspiracy, and that this requires 'pursuing a *common aim* which is contrary to the law'.[105] Advocate General Darmon expressed a similar view, finding the origin of the concept of concerted practice in the US antitrust concept of concerted action.[106] As expressed in

[101] Joined Cases C-89/85, C-104/85, C-114/85, C-116/85, C-117/85 and C-125/85 to C-129/85 *A. Ahlström Osakeyhtiö v Commission* [1993] ECR I-1307, AG Opinion para 166.

[102] The second limb is important because no legal consequences turn on whether competition is restricted by agreement or by concerted practice: Case C-49/92 P *Commission v Anic Partecipazioni SpA* [1999] ECR I-4125, para 132–133 and Whish (2003: 94–96).

[103] Joined Cases C-89/85, C-104/85, C-114/85, C-116/85, C-117/85 and C-125/85 to C-129/85 *A. Ahlström Osakeyhtiö v Commission* [1993] ECR I-1307, AG Opinion para 165, Steindorff (1972: 508), Dashwood *et al.* (1976: 481), Mann (1973: 38–41), and Gijlstra and Murphy (1977: 58–60).

[104] Honig *et al.* (1963: 10), emphasis added. Also Gleiss and Hirsch (1981: 53), Oberdorfer and Gleiss (1963: 8), and Wessely (2001: 746).

[105] Case 48/69 *Imperial Chemical Industries v Commission* [1972] ECR 619, 669, emphasis added. Korah (1973: 222) criticizes the AG's reading of the US position, contending that 'he does not appear to consider the dicta in the light of the facts'.

[106] Joined Cases C-89/85, C-104/85, C-114/85, C-116/85, C-117/85 and C-125/85 to C-129/85 *A. Ahlström Osakeyhtiö v Commission* [1993] ECR I-1307, AG Opinion para 166, n 71. For support of the AG's view on the origin of the term 'concerted practice': Joined Cases

US antitrust the concepts 'necessarily presuppose an "*agreement*", that is to say a *meeting of minds*'.[107] For some, a concerted practice exists when certain conduct 'can only be explained by a *common intention* on the part of those undertakings'.[108] For Mann:

It could not be otherwise, for Article [81] puts concerted practices on the same level as agreements . . . so that established rules of treaty interpretation render it impossible to attribute a meaning to the term "concerted practices" which is not *ejusdem generis* and would render some form of consensus redundant.[109]

(a) The relationship between agreement and concerted practice Since concerted practice is included to capture conduct not already caught by 'agreement', and since common intention *is* caught by agreement, there must be something distinctive about common intention in concerted practice. An idea underlying some cases and commentary is that common intention in agreement is expressed in a manner that gives rise to (legal) obligation, whereas common intention in concerted practice is not so expressed.[110]

Whether common intention must be expressed in a manner giving rise to (legal) obligation in order to be captured as an 'agreement' within Article 81 EC was first considered by the Court in the *Chemiefarma* action to annul the Commission's *Quinine* decision. Cooperation had existed between quinine producers since 1913; in 1958, eight of the leading producers

C-89/85, C-104/85, C-114/85, C-116/85, C-117/85 and C-125/85 to C-129/85 *A. Ahlström Osakeyhtiö v Commission* [1993] ECR I-1307, AG Opinion para 166, n 71, Case T-1/89 *Rhône-Poulenc SA v Commission* [1991] ECR II-867, AG Opinion 927–929, Deringer (1968: 11–12), Gleiss and Hirsch (1981: 52), Goyder (2003: 72), and Oberdorfer and Gleiss (1963: 8). Guerrin and Kyriazis (1993: 778–779) identify the origin in Article 419 of the French Penal Code of 1810. On the concept as it developed in US antitrust: Kovacic (1993: 22–25).

[107] Joined Cases C-89/85, C-104/85, C-114/85, C-116/85, C-117/85 and C-125/85 to C-129/85 *A. Ahlström Osakeyhtiö v Commission* [1993] ECR I-1307, AG Opinion para 166, n 71, emphasis added. Also Guerrin and Kyriazis (1993: 790, 792–794) and Rose (1993: para 2-034).

[108] Case 48/69 *Imperial Chemical Industries v Commission* [1972] ECR 619, para 113. Compare with the discussion of 'terms' in Black (2003b: 408–409).

[109] Mann (1973: 36).

[110] Joined Cases C-89/85, C-104/85, C-114/85, C-116/85, C-117/85 and C-125/85 to C-129/85 *A. Ahlström Osakeyhtiö v Commission* [1993] ECR I-1307, AG Opinion para 166–167 and n 71, *Re Cartel in Quinine* [1969] CMLR D41, D59, Case 41/69 *ACF Chemiefarma NV v Commission of the European Communities* [1970] ECR 661, 716–718, Guerrin and Kyriazis (1993: 790, 792–794), Joshua and Jordan (2004: 658), and Rose (1993: para 2-034). In Case 51/75 *EMI Records v CBS United Kingdom Limited* [1976] ECR 811, para 30–31, it is held that if a legal agreement is terminated but continues to influence future conduct, conduct after the termination of the agreement constitutes a concerted practice. See Van den Bergh and Camesasca (2001: 176, 181–182), Martin (2001: 64), Neven *et al.* (1998: 62–63), and Faull and Nikpay (1999: para 2.30).

entered into a formal, legally binding, written agreement.[111] Following an objection raised by the Bundeskartellamt over the compatibility of the agreement with EC law and disagreement between the contracting parties, the 1958 agreement was abandoned. Some of the quinine producers entered into a new formal, legally binding, written agreement.[112] This new, formal, legally binding, written agreement did not apply to activities within the EC.[113] However, it was agreed that the terms from the formal, legally binding, written agreement would apply within the EC, but that the document extending the terms to the EC would remain unsigned to prevent it creating legal obligations, though it was to be enforceable by arbitration.[114]

The Commission felt that both the signed and unsigned documents evidenced agreement. Although said to be non-binding, the unsigned document 'expressly laid down, in writing, binding provisions . . . to which the parties could be held through arbitration proceedings'.[115] The unsigned agreement 'had to be respected in spite of its non-obligatory character'.[116] The terms of the unsigned document were in fact adhered to.[117] In his analysis, Advocate General Gand felt that 'evidence of the *binding nature* of the gentlemen's agreement is clearly expressed in the provision stipulating that a breach of the gentlemen's agreement *ipso facto* constitutes a breach of the [formal] agreement'.[118] The Court considered the gentlemen's agreement as a term of the legally binding export agreement, so that the gentlemen's agreement was not without legal effect.[119] Thus, although *Chemiefarma* is now cited as authority for the proposition that an 'agreement' does not need to be legally binding for the purpose of Article 81 EC, it is clear that the Commission, the Advocate General, and the Court felt that the 'gentlemen's agreement' *was* (legally) binding. This appears to be confirmed by Advocate General Mayras writing in a subsequent case that

[111] *Re Cartel in Quinine* [1969] CMLR D41, D42–D44 and Case 41/69 *ACF Chemiefarma NV v Commission of the European Communities* [1970] ECR 661, para 1.

[112] *Re Cartel in Quinine* [1969] CMLR D41, D45 and Case 41/69 *ACF Chemiefarma NV v Commission of the European Communities* [1970] ECR 661, para 2–4.

[113] *Re Cartel in Quinine* [1969] CMLR D41, D46.

[114] *Re Cartel in Quinine* [1969] CMLR D41, D46–D49 and Case 41/69 *ACF Chemiefarma NV v Commission of the European Communities* [1970] ECR 661, para 5–7.

[115] *Re Cartel in Quinine* [1969] CMLR D41, D58.

[116] *Re Cartel in Quinine* [1969] CMLR D41, D54.

[117] *Re Cartel in Quinine* [1969] CMLR D41, D49–D50, D59.

[118] Case 41/69 *ACF Chemiefarma NV v Commission of the European Communities* [1970] ECR 661, 715, emphasis added, also 714.

[119] Case 41/69 *ACF Chemiefarma NV v Commission of the European Communities* [1970] ECR 661, para 111–113.

'[u]p till now, the Court has only had to consider the application or interpretation of Article [81] in relation to *agreements* between undertakings, that is to say *contracts* concluded between producers or between producers and vendors'.[120] In *FEDETAB*, the seventeenth submission was that '[o]nly a binding *contract* under national law, made between two or more parties, may be described as an agreement'.[121] Advocate General Reischl considered whether, '[p]roperly interpreted however, agreements within the meaning of Article [81(1)] of the [EC] Treaty are but *contracts* under national law' and tentatively suggested that 'the view *may* be taken that the concept of "agreement" within the meaning of Article [81(1)] is wider than that of contract in civil law'.[122] *Chemiefarma* was too equivocal to displace the view that '[a]greements are bilateral or multilateral understandings in which at least one partner is *legally bound*', which persists until more definitively rejected in *Sandoz*.[123]

The idea that 'agreement' captures only binding expressions of common intention leaves a gap to be filled by a conception of concerted practice covering expressions of common intention lacking binding force.[124] In an action to annul the Commission's *Dyestuffs* decision, the Court states that concerned practice involves 'a form of coordination' that has yet to reach the stage of 'agreement properly so-called'.[125] The conception of concerted practice discussed seems to take 'agreement properly so-called' to be contract; a concerted practice 'does not have all the elements of a contract'.[126]

[120] Case 48/69 *Imperial Chemical Industries v Commission* [1972] ECR 619, 669, 'agreements' is emphasized in the original and 'contracts' is additionally emphasized here.

[121] Joined Cases 209 to 215 and 218/78 *Heintz Van Landewyck SARL and Others v Commission (FEDETAB)* [1980] ECR 3125, 3221, emphasis in original.

[122] Joined Cases 209 to 215 and 218/78 *Heintz Van Landewyck SARL and Others v Commission (FEDETAB)* [1980] ECR 3125, 3310, emphasis added. The Court did not have to consider whether agreement captures more that binding expressions of common intention because the conduct in question was binding: 3226, para 85–86.

[123] Gleiss and Hirsch (1981: 49), emphasis added. See C-277/87 *Sandoz Prodotti Farmaceutici SpA v Commission* [1990] ECR I-45, Section II.B above, Deringer (1968: para 414), Oberdorfer and Gleiss (1963: 6), and Korah (2004: 56). Under the ECMR an agreement exists when it 'cannot be rescinded unilaterally and [that] intends to create a legal relationship upon which each party can rely': Levy (2003: para 17.03(2)).

[124] The significance does not appear to be generally recognized, so that in Case T-1/89 *Rhône-Poulenc SA v Commission* [1991] ECR-II 867, AG Opinion 926, it is written that 'whether one chooses to regard non-binding arrangements as agreements within the meaning of Article [81] or to reserve the term concerted practices for practical cooperation which has not been given formal expression, it is the whole gamut of anti-competitive arrangements which is caught by Article [81].'

[125] Case 48/69 *Imperial Chemical Industries v Commission* [1972] ECR 619, para 64.

[126] Case 48/69 *Imperial Chemical Industries v Commission* [1972] ECR 619, para 64, 65.

Advocate General Mayras considered that agreement and concerted practice were distinct because, whilst both involve common intention, the latter 'is not set out in a document which has the purpose of determining the respective obligations of the parties'.[127] Both the fourth and fifth edition of *Bellamy & Child* express the view that 'if the parties combine informally to restrict competition, even if the arrangements are *binding neither legally nor morally*, they may give rise to a "concerted practice"'.[128] On this view 'the only difference between concerted practices and agreements lies in the impossibility of enforcing a concerted practice in court'.[129] Thus, '[i]n general it can be said that agreements . . . obligate the parties, as a matter of law or fact, to maintain a certain conduct, whereas concerted practices do not have this effect.'[130]

(b) Evidence of non-binding common intention Under this conception, the distinction between concerted practice and agreement lies in how common intention is expressed and thus evidenced. The inclusion of the concept of concerted practice in Article 81 EC 'enables the authorities to intervene where there is no direct evidence of an agreement, but where circumstances permit to presume its existence with certainty'.[131] When there is an overwhelmingly strong conviction that conduct is directed by some common intention, the basis of that overwhelmingly strong conviction can be called to evidence common intention. There is perhaps no clearer statement of this position than that of Advocate General Mayras, writing that:

> if it is accepted that *a concerted practice in fact conceals an agreement and at the same time reveals it through the appearance of some coordinated conduct*, there is to my mind no doubt that in making a separate 'category' for concerted practices, the authors of the Treaty have intended to avoid the possibility of undertakings' evading the prohibitions of Article [81] . . . by *so conducting their affairs as not to leave any written document which might be called an agreement*.
>
> Such an interpretation, which takes practical account of the distinction made in Article [81], is of obvious interest as regards evidence for the existence of a concerted practice which, even though it implies that *the will of the participating*

[127] Case 48/69 *Imperial Chemical Industries v Commission* [1972] ECR 619, AG Opinion 678, also 673.

[128] Roth QC (2001: para 2-016, emphasis added) and Rose (1993). Also Steindorff (1972: 508). Compare Joshua and Jordan (2004: 660).

[129] Wielen (1971: 88).

[130] Deringer (1968: para 115). Also Gleiss and Hirsch (1981: 50). In Roth QC (2001: para 2-015, n 47) it is written that 'an "informal agreement" may also be a concerted practice'.

[131] Waelbroeck and Frignani (1999: 131). Also Black (2003a: 71–75), Furse (2004: 146–152), Martin (2001: 59–64), Neven *et al.* (1998: 45–78), and Van den Bergh and Camesasca (2001: 168).

undertakings is somehow apparent, nevertheless cannot be sought using the same methods as for proof of an express agreement.[132]

The idea of concerted practice as a means of evidencing common intention posits a continuum along which evidence is more or less probative in showing common intention.[133] This conception fits with the view that the forms of collusion enumerated in Article 81 EC have 'the same nature and are only distinguishable from each other by their intensity and the forms in which they manifest themselves'.[134] The continuum conception is illustrated in Figure 4.1. As we move away from the top left hand corner, the evidence is less likely to establish agreement and more likely to establish concerted practice. Written and parol evidence of common intention establishes agreement; evidence of common intention inferred from conduct establishes concerted practice.[135]

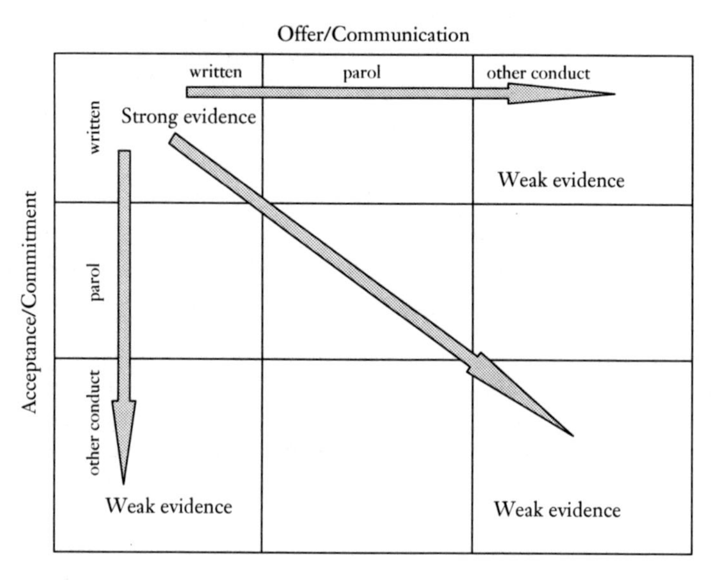

Figure 4.1 Evidence of Common Intention.

[132] Case 48/69 *Imperial Chemical Industries v Commission* [1972] ECR 619, AG Opinion 671, emphasis added, also 673.

[133] Compare Faull and Nikpay (1999: para 2.54), Whish (2003: 94–96) and Joined Cases 209 to 215 and 218/78 *Heintz Van Landewyck Sarl and Others v Commission (FEDETAB)* [1980] ECR 3125, AG Opinion 3310.

[134] Case C-49/92 P *Commission v Anic Partecipazioni SpA* [1999] ECR I-4125, para 131. Also Case T-41/96 *Bayer AG v Commission* [2000] ECR II-3383, para 69.

[135] Kovacic (1993: 19–21) takes a similar approach when analysing agreement in the US context. However, the distinction drawn along his continuum is between express and tacit agreement.

Three factors enable the inference of common intention to be made from conduct. First, there must be communication.[136] Communication may involve physical meeting (contact), but any form of communication is sufficient. Second, there must be conduct subsequent to communication. If agreement is to be inferred from conduct, logically, conduct must exist: 'the existence of a concerted practice cannot be established if the parties do not act on their mutual arrangements'.[137] In other words, concertation must have been put into practice. Third, once communication is established, the question is whether it is possible to infer that consensus was reached from the subsequent behaviour of the undertakings concerned.[138] The inference of common intention from conduct cannot be made unless it is feasible that consensus was reached during communication and consensus is the only plausible explanation for the conduct.[139] As the Court has made clear, 'conduct cannot be regarded as furnishing proof of concertation unless concertation constitutes the *only* plausible explanation for such conduct'.[140] There is a view that we can only have an overwhelmingly strong conviction that conduct is guided by common intention reached during contact if contact is followed by 'identical action'.[141]

The inference of common intention from conduct (practice) subsequent to communication can be seen in *Dyestuffs*. Between 7 and 20 January 1964,

[136] Antunes (1991: 62).

[137] Antunes (1991: 65). On the debate over whether concertation needs to be put into practice: Joined Cases C-89/85, C-104/85, C-114/85, C-116/85, C-117/85 and C-125/85 to C-129/85 *A. Ahlström Osakeyhtiö v Commission* [1993] ECR I-1307, AG Opinion para 184–188, Case T-1/89 *Rhône-Poulenc SA v Commission* [1991] ECR II-867, AG Opinion 927–945, Case T-2/89 *Petrofina v Commission* [1991] ECR II-1087, para 197–216, Case C-49/92 P *Commission v Anic Partecipazioni SpA* [1999] ECR I-4125, AG Opinion para 12–30, Antunes (1991: 64–70), Baardman (1971: 90), van Gerven and Navarro Varona (1994: 597–599), Waelbroeck and Frignani (1999: para 137), Wessely (2001: 743–747, 751–760), Whish (2000: 224), and Wielen (1971: 87–88).

[138] Joined Cases 40 to 48, 50, 54 to 56, 111, 113 and 114–73 *Coöperatieve Vereniging 'Suiker Unie' Ua v Commission* [1975] ECR 1663, AG Opinion 2060 and Guerrin and Kyriazis (1993: 804–812).

[139] Posner (2001: 60–79). Collusion is more feasible and plausible in industries populated by few undertakings, in which the products are homogenous: Stigler (1983b), Carlton and Perloff (2000: 136–141), Van den Bergh and Camesasca (2001: 170–176), Phlips (1995: 23–38), Tirole (1988: 240–243), and Scherer and Ross (1990: 277–294).

[140] Joined Cases C-89/85, C-104/85, C-114/85, C-116/85, C-117/85 and C-125/85 to C-129/85 *A. Ahlström Osakeyhtiö v Commission* [1993] ECR I-1307, para 71, emphasis added, also AG Opinion para 195–197, and Soames (1996: 24–29).

[141] Mann (1973: 36, emphasis added). Also Case 48/69 *Imperial Chemical Industries v Commission* [1972] ECR 619, AG Opinion 671, judgment para 66, Case 41/69 *ACF Chemiefarma NV v Commission of the European Communities* [1970] ECR 661, AG Opinion 714–715, and Wessely (2001: 756). In US antitrust, it is said that there must be 'plus factors': Kovacic (1993: 31–55).

ten producers of aniline dyes increased their prices by 15 per cent, then by 10 per cent on 1 January 1965, and in October 1967 by between 8 per cent and 12 per cent.[142] The Commission pointed out that the increases were applied to the same products, even though the 10 firms each produced between 1,500 and 3,500 of some 6,000 dyes.[143] The head offices of a number of firms sent instructions to their subsidiaries to increase the price by telex or telegram within a three-hour window on 9 January 1964.[144] The instructions to increase the price contained 'great similarity in drafting, to the extent of containing identical phrases'.[145] The Commission concluded that '[i]t is not conceivable that *without detailed prior agreement* the principle producers... should several times increase by identical percentages the prices'.[146]

The inference of common intention from conduct (practice) subsequent to communication can also be seen in *Suiker Unie*. A meeting of French, German, Italian, Belgian and Dutch sugar producers was held at Munich in May 1968.[147] French and Belgian sugar producers met in Paris on 29 July 1969 and in Genoa on 11 September 1970.[148] Subsequently there was parallel conduct: sugar producers would only export sugar within the Common Market to other sugar producers, so that industrial consumers could only obtain supplies from their national producers.[149] The question was whether it is possible to infer that consensus was reached at the various meetings in the absence of written or parol evidence of common intention. The inference can be made if conduct subsequent to communication

[142] Case 48/69 *Imperial Chemical Industries v Commission* [1972] ECR 619, para 1–6, 83–87.

[143] *Re Cartel in Aniline Dyes* [1969] CMLR D23, para 7, Korah (1973: 220–221), and Case 48/69 *Imperial Chemical Industries v Commission* [1972] ECR 619, AG Opinion 674.

[144] *Re Cartel in Aniline Dyes* [1969] CMLR D23, para 7.

[145] *Re Cartel in Aniline Dyes* [1969] CMLR D23, para 8. For discussion: Joshua and Jordan (2004: 658–660).

[146] *Re Cartel in Aniline Dyes* [1969] CMLR D23, para 7, emphasis added. Also Case 48/69 *Imperial Chemical Industries v Commission* [1972] ECR 619, para 54, Case 49/69 *BASF v Commission* [1972] ECR 713, para 35, Mann (1973: 39–41). Dyestuffs had been heavily cartelized in Europe since at least 1927: Stocking and Watkins (1946: 466–517, particularly 505–511).

[147] Joined Cases 40 to 48, 50, 54 to 56, 111, 113 and 114–73 *Coöperatieve Vereniging 'Suiker Unie' Ua v Commission* [1975] ECR 1663, AG Opinion 2058.

[148] Joined Cases 40 to 48, 50, 54 to 56, 111, 113 and 114–73 *Coöperatieve Vereniging 'Suiker Unie' Ua v Commission* [1975] ECR 1663, AG Opinion 2068. For the Italian market, the parties did not dispute the existence of agreement, though there was no contract; instead, they said there was no competition to restrict: para 29–73 and Dashwood *et al.* (1976: 482).

[149] Joined Cases 40 to 48, 50, 54 to 56, 111, 113 and 114–73 *Coöperatieve Vereniging 'Suiker Unie' Ua v Commission* [1975] ECR 1663, para 254–257, 318–326, AG Opinion 2058. There was also much documentary evidence: para 131–155, 200–203, 246–253, 258–270, 275, 327–331, AG Opinion 2058, and Dashwood *et al.* (1976: 485).

'cannot be explained other than by concertation'.[150] Advocate General Mayras felt that the uniform practices 'can only be explained by a co-ordinated policy, the principles of which, laid down by the most important producers, were intended to maintain the partitioning of national markets'.[151]

There are also failures to infer common intention from conduct (practice) subsequent to communication. In *Wood pulp*, over one hundred firms from eighteen countries supplied around eight hundred Community paper manufacturers. Wood pulp producers did not sell directly to the paper manufacturers; the goods were marketed through a system of agents or cooperatives.[152] To ensure a stable supply of the right mix of pulp, paper manufacturers entered long term supply contracts with pulp producers. Producers guaranteed supply of a minimum quantity at a price not to exceed a maximum, which was announced each quarter.[153] The Commission decided that forty wood pulp producers and three of their trade associations had engaged in concerted practices.[154]

It was alleged that the producers had concerted (a) on the prices that they would announce, and (b) on the prices they would actually charge.[155] Pulp prices were announced simultaneously; were always quoted in US dollars; included the cost of delivery; were identical, and were adhered to.[156] The Commission considered that this conduct could only be explained by the existence of an agreement on price that must have been

[150] Joined Cases C-89/85, C-104/85, C-114/85, C-116/85, C-117/85 and C-125/85 to C-129/85 *A. Ahlström Osakeyhtiö v Commission* [1993] ECR I-1307, para 184, also para 183–198, Turner 1962: (659–660, 672–673), Neven *et al.* (1998: 59, 62–76), and van Gerven and Navarro Varona (1994: 601–604). Compare Kovacic (1993: 55–57).

[151] Joined Cases 40 to 48, 50, 54 to 56, 111, 113 and 114–73 *Coöperatieve Vereniging 'Suiker Unie' Ua v Commission* [1975] ECR 1663, AG Opinion 2059; upheld by the court at para 167–192.

[152] Joined Cases C-89/85, C-104/85, C-114/85, C-116/85, C-117/85 and C-125/85 to C-129/85 *A. Ahlström Osakeyhtiö v Commission* [1993] ECR I-1307, 1321–1322. The cooperatives are described in the decision as associations of undertakings, but they sell the members' pulp on their own account. On *Wood pulp*: van Gerven and Navarro Varona (1994) and Soames (1996: 28–29).

[153] Joined Cases C-89/85, C-104/85, C-114/85, C-116/85, C-117/85 and C-125/85 to C-129/85 *A. Ahlström Osakeyhtiö v Commission* [1993] ECR I-1307, 1323.

[154] Joined Cases C-89/85, C-104/85, C-114/85, C-116/85, C-117/85 and C-125/85 to C-129/85 *A. Ahlström Osakeyhtiö v Commission* [1993] ECR I-1307, 1320.

[155] Joined Cases C-89/85, C-104/85, C-114/85, C-116/85, C-117/85 and C-125/85 to C-129/85 *A. Ahlström Osakeyhtiö v Commission* [1993] ECR I-1307, 1324.

[156] Joined Cases C-89/85, C-104/85, C-114/85, C-116/85, C-117/85 and C-125/85 to C-129/85 *A. Ahlström Osakeyhtiö v Commission* [1993] ECR I-1307, 1323–1326, 1415–1416, contested at 1347–1346.

reached at some stage prior.[157] The claim was that the conduct was so contrary to what would normally be expected that we could infer the existence of a price-fixing agreement from the conduct—the conduct could not be explained absent such agreement.[158] The parties argued, and the Advocate General and Court accepted, that there were many reasons why prices were announced quarterly; simultaneously; quoted in US dollars; included the cost of delivery; and were identical: a prior agreement was not the only plausible or credible explanation for the conduct.[159]

(c) Concerted practice as common intention rejected It can be doubted whether concerted practice really is simply a way to evidence common intention. The idea that 'concerted practices are a pale or diluted form of agreement' has been challenged.[160] First, it is clear that common intention is not required; in *Suiker Unie* it is confirmed that concerted practices 'in no way require the working out of an actual plan.'[161] Contemporary commentators considered that this 'establishes beyond any doubt' that a concerted practice 'does not require a meeting of the minds in the sense of agreement, however informal, as to the course of future conduct'.[162] The Court is clear that 'a concerted practice does not have all the elements of a contract'.[163] The view of concerted practice as a way to evidence common intention reads this as saying that, unlike contract, common intention need not be expressed in any particular way. However, it can also be read as saying that a concerted practice does not require common intention.[164]

[157] Joined Cases C-89/85, C-104/85, C-114/85, C-116/85, C-117/85 and C-125/85 to C-129/85 *A. Ahlström Osakeyhtiö v Commission* [1993] ECR I-1307, 1410–1411, 1414–1415, AG Opinion para 241–243, 319–327, judgment para 57. There was documentary evidence that the undertakings met and that prices were discussed. However, the parties denied that meetings took place, and Court ruled the documents inadmissible: 1324–1325, 1329, 1346, judgment para 66–69, AG Opinion para 454–476.

[158] Joined Cases C-89/85, C-104/85, C-114/85, C-116/85, C-117/85 and C-125/85 to C-129/85 *A. Ahlström Osakeyhtiö v Commission* [1993] ECR I-1307, 1438–1441, AG Opinion para 279–296.

[159] Joined Cases C-89/85, C-104/85, C-114/85, C-116/85, C-117/85 and C-125/85 to C-129/85 *A. Ahlström Osakeyhtiö v Commission* [1993] ECR I-1307, para 70–127, AG Opinion para 279–296, 328–434.

[160] Black (2003c: 106–107).

[161] Joined Cases 40 to 48, 50, 54 to 56, 111, 113 and 114–73 *Coöperatieve Vereniging 'Suiker Unie' Ua v Commission* [1975] ECR 1663, para 173.

[162] Dashwood *et al.* (1976: 486). Also Case T-1/89 *Rhône-Poulenc SA v Commission* [1991] ECR II-867, AG Opinion 930–931, Gijlstra and Murphy (1977: 58–59), Waelbroeck and Frignani (1999: 135), and Antunes (1991: 67).

[163] Case 48/69 *Imperial Chemical Industries v Commission* [1972] ECR 619, para 65.

[164] Compare Case T-41/96 *Bayer AG v Commission* [2000] ECR II-3383, para 65.

Second, it seems strange for the EC Treaty to provide a means of establishing common intention with evidence that is too weak to establish common intention as agreement. Common intention is not a continuum concept; either there is a concurrence of wills, or there is not. The evidence used to establish common intention may fit along a continuum ranging from strong to weak, but the evidence is either sufficiently compelling to establish common intention or it is not.[165]

Finally, and perhaps most importantly, the conception of concerted practice as common intention captures no more activity than is already caught within the meaning of agreement. The concept of concerted practice in the EC Treaty is said to originate from US case law interpreting the Sherman Act conception of conspiracy.[166] The US conception indeed exists in part to resolve 'the evidentiary burden of proving illegal collusion where only circumstantial evidence, and not direct evidence of meetings and agreements, can be offered'.[167] However, in Article 81 EC Treaty, 'a full "agreement" can still be proved entirely by circumstantial evidence'.[168] Concerted practice as common intention does not provide 'an independent scope which may be described as covering cooperation which is not an agreement'.[169] The conception of concerted practice as common intention fails to recognize the concern that:

> [t]o give [concerted practices] so limited and so narrow a meaning that it would be reduced to nothing more than a particular application of the concept of agreement would, quite undoubtedly, be contrary to a general principle of interpretation . . . [that] full effect is to be given to each of the provisions of the Treaty and they are all to be given full scope.[170]

2. *Reduction of uncertainty*

'Agreement' captures expressions of common intention that are shown to exist by evidence of communication (offer) and commitment (acceptance), be it written, parol, or inferred from conduct. It is acknowledged that the purpose of concerted practice is to capture forms of conduct not caught by

[165] Compare Gellhorn *et al.* (2004: 269–270).

[166] See references cited at note 105–106 above.

[167] Scherer and Ross (1990: 339). Also Ross 1993: (158–182), Kaysen (1951), and Posner (1969: 1578–1587).

[168] Joshua and Jordan (2004: 660).

[169] Case T-1/89 *Rhône-Poulenc SA v Commission* [1991] ECR II-867, AG Opinion 929. Also Case C-49/92 P *Commission v Anic Partecipazioni SpA* [1999] ECR I-4125, para 112.

[170] Case 48/69 *Imperial Chemical Industries v Commission* [1972] ECR 619, AG Opinion 671.

'agreement'.[171] Concerted practice must also have something in common with agreement so that both can be seen as forms of collusion.[172] It may be that what agreement and concerted practice share is not the constituent feature of common intention, but that undertakings engage in both agreements and concerted practices for the same reason. Rather than ask what agreements and concerted practices are, we should consider why undertakings enter into agreements or engage in concerted practices.

For Fried, the purpose of agreement is to 'tie down the *future*'.[173] By providing *ex post* enforcement, the law encourages *ex ante* reliance on agreement.[174] Legally enforceable agreements reduce uncertainty as to the future.[175] Other than agreement, undertakings may interact in various ways and engage in various strategies to reduce uncertainty as to the future. An alternative conception of concerted practice, distinct from 'agreement' because common intention is not required, is revealed if we focus on techniques used to reduce uncertainty as to future conduct of others.[176]

Focusing on the potential of various strategies to reduce uncertainty about the future conduct of others provides a framework in which the jurisprudence of the Court can be understood. In *Dyestuffs*, a concerted practice is said to involve 'knowingly substitut[ing] practical cooperation . . . for the *risks of competition*'.[177] Risk and uncertainty go together. The Court was satisfied that a concerted practice existed once it was clear that the:

> undertakings *eliminated all uncertainty* between them as to their future conduct and, in doing so, also *eliminated a large part of the risk* usually inherent in any independent change of conduct on one or several markets.[178]

[171] Faull and Nikpay (1999: para 2.45), Rose (1993: para 2-034), Goyder (2003: 71–72), and Whish (2003: 99).

[172] See note 102 above.

[173] Fried (1981: 45, emphasis added, also 8–21), Beatson (2002: 2–3), Kimel (2003: 7–87), Shavell (2004: 297), Smith (2004: 71–78), and Treitel (2004: 5–6).

[174] Bayles (1987: 148–149), Blair and Kaserman (1985: 409–413), Carlton and Perloff (2000: 405), Dobson and Waterson (1996: 18), Schwartz and Scott (2003: 550–568), Smith (2004: 108–136), Tirole (1988: 185), Williamson (1974: 1450–1463), Williamson (1977: 723–726), Williamson (1979a: 239–242), and Williamson (1981: 55).

[175] Bayles (1987: 143–146), Buckley (2005: 35–50), Motta (2004: 141), Shavell (2004: 293–296), and Smith (2004: 110–113).

[176] Compare: *Argos Limited and Littlewoods Limited v Office of Fair Trading* [2004] CAT 24, para 659–664 and Faull and Nikpay (1999: para 2.45–2.55).

[177] Case 48/69 *Imperial Chemical Industries v Commission* [1972] ECR 619, para 64, emphasis added.

[178] Case 48/69 *Imperial Chemical Industries v Commission* [1972] ECR 619, para 101, emphasis added. Compare Joined Cases 40 to 48, 50, 54 to 56, 111, 113 and 114–73 *Coöperatieve Vereniging 'Suiker Unie' Ua v Commission* [1975] ECR 1663, para 180. On the uncertainty that existed: Case

Rather than face the uncertainty of competition, concerted practices put undertakings 'in a position to assess the market situation with *considerably more certainty* and to act accordingly'.[179] Concerted practice as conduct reducing uncertainty is seemingly confirmed in *Suiker Unie*, which remains the leading case on the concept of concerted practice.[180]

The criterion of reduced uncertainty is an appropriate one with which to draw the boundary of Article 81 EC. Competition requires uncertainty; the risks of competition are that an undertaking does not know how competitors, trading partners, and customers will act in the future; it is this uncertain environment that provides the impetus to engage in the competitive struggle. In *Suiker Unie*, the need for uncertainty and the competitive risk that this entails are recognized when the Court requires 'that each economic operator must determine independently the policy which he intends to adopt on the Common Market'.[181] In *Tate & Lyle*, the CFI recognized 'that *uncertainty* as to the pricing policies which the other operators intend to practise in the future constitutes the main stimulus to competition'.[182] A concerted practice enables undertakings to 'act with greater knowledge and more or less justified expectations about other undertakings than they should have had and normally would have'.[183] Reduced uncertainty provides comfort for those who 'prefer . . . the security of keeping what they already ha[ve] to the chance of winning more, when this entail[s] the risk of total loss'.[184] Faull and Nikpay thus write that any 'direct or indirect contact which is designed or has the effect of *reducing uncertainty about their future conduct* is likely to be found to be a concerted practice'.[185]

48/69 *Imperial Chemical Industries v Commission* [1972] ECR 619, para 69–71, 107–108, AG Opinion 674–675.

[179] Case T-1/89 *Rhône-Poulenc SA v Commission* [1991] ECR II-867, AG Opinion 941, emphasis added.

[180] Joined Cases 40 to 48, 50, 54 to 56, 111, 113 and 114–73 *Coöperatieve Vereniging 'Suiker Unie' Ua v Commission* [1975] ECR 1663, para 26, Case C-49/92 P *Commission v Anic Partecipazioni SpA* [1999] ECR I-4125, para 41, 115–117 and Case C-199/92 P *Hüls AG v Commission* [1999] ECR I-4287, para 158–160.

[181] Joined Cases 40 to 48, 50, 54 to 56, 111, 113 and 114–73 *Coöperatieve Vereniging 'Suiker Unie' Ua v Commission* [1975] ECR 1663, para 173. Also Case C-49/92 P *Commission v Anic Partecipazioni SpA* [1999] ECR I-4125, AG Opinion para 25 and Turner (1962: 673).

[182] Joined Cases T-202/98, T-204/98 and T-207/98 *Tate & Lyle PLC, British Sugar PLC, Napier Brown & Co. Ltd v Commission* [2001] ECR II-2035, para 46, emphasis added.

[183] Case T-1/89 *Rhône-Poulenc SA v Commission* [1991] ECR II-867, AG Opinion 942.

[184] Deringer (1962: 1).

[185] Faull and Nikpay (1999: para 2.49, emphasis added). In Case C-49/92 P *Commission v Anic Partecipazioni SpA* [1999] ECR I-4125, AG Opinion para 34–36, the AG questions whether

B. *Proof of concerted practice*

It is argued that the concerted practice as a means of evidencing common intention captures no more than is already caught by the conception of agreement discussed in Section II above. It is also argued that it is more illuminating to consider *why* undertakings engage in concerted practices than to ask *what* concerted practices are. What is then revealed is a conception of collusion focusing on reductions of uncertainty as to the future conduct of others; both agreements and concerted practices can be used to reduce uncertainty. If reduction of uncertainty as to the future conduct of others is, or is to become, an accepted and acceptable conception of concerted practice, it must be possible to prove that a reduction of uncertainty has occurred. Communication reduces uncertainty and evidence of communication seems sufficient to establish a concerted practice.[186]

Communication falls within the boundary of Article 81 EC as it 'enables each of the parties to determine its policy without having to be subject to the risks of competition'.[187] Even though a commitment as to how the information will be used is absent, undertakings receiving information cannot fail to consider it when determining their future conduct; conduct is inevitably influenced.[188] In *Anic*, the Court established a strong presumption of influence, writing that:

> subject to proof to the contrary . . . there must be a presumption that the undertakings participating in concerting arrangements and remaining active on the market take account of the information exchanged with their competitors when determining their conduct on that market.[189]

' "lessening of uncertainty" as to the future' is sufficient and asks whether 'certainty' ought to be created.

[186] Joined Cases C-89/85, C-104/85, C-114/85, C-116/85, C-117/85 and C-125/85 to C-129/85 *A. Ahlström Osakeyhtiö v Commission* [1993] ECR I-1307, AG Opinion para 172, Antunes (1991: 61–62, 67), Faull and Nikpay (1999: para 2.49), Kuhn (2001: 179–187), and Waelbroeck and Frignani (1999: 135).

[187] Antunes (1991: 66). Also Wessely (2001: 756, 759).

[188] Joined Cases 40 to 48, 50, 54 to 56, 111, 113 and 114–73 *Coöperatieve Vereniging 'Suiker Unie' Ua v Commission* [1975] ECR 1663, para 174, Case T-1/89 *Rhône-Poulenc SA v Commission* [1991] ECR II-867, para 122–124, Joined Cases T-202/98, T-204/98 and T-207/98 *Tate & Lyle PLC, British Sugar PLC, Napier Brown & Co. Ltd v Commission* [2001] ECR II-2035, para 34, 58, Case CP/0871/01 *Price-Fixing of Replica Football Kit* [2003] 1 August, para 311, 406, 436, Case 1021/1/1/03 and 1022/1/1/03 *JJB Sports PLC v Office of Fair Trading; Allsports Limited v Office of Fair Trading* [2004] CAT 17, para 877–878, Case 1032/1/1/04 *Apex Asphalt and Paving Co Limited v Office of Fair Trading* [2005] CAT 4, para 179, Case 1033/1/1/04 *Richard W Price (Roofing Contractors) Limited v Office of Fair Trading* [2005] CAT 5, para 50–51 and Antunes (1991: 66–67).

[189] Case C-49/92 P *Commission v Anic Partecipazioni SpA* [1999] ECR I-4125, para 121. Also Joined Cases C-89/85, C-104/85, C-114/85, C-116/85, C-117/85 and C-125/85 to C-129/85

This view is shared by Judge Vesterdorf, writing as Advocate General. Once there is communication:

undertakings will then necessarily, and normally unavoidably, act on the market in the light of the knowledge and on the basis of the discussions which have taken place . . . They will negotiate with their customers and arrange their production and so forth possessing a different body of knowledge and being in a different state of awareness than if they had only their own experience, general knowledge and perception of the market to rely on.[190]

What constitutes communication remains unclear.[191] It is clear that an *agreement* to exchange information can be scrutinized as an 'agreement' within the meaning of Article 81 EC.[192] Even when not done pursuant to an agreement, exchanging information is communication that reduces uncertainty and evidences concerted practice.[193] Beyond this, things become fuzzy. One view is that evidence that information has been requested, or simply accepted, shows communication:

A Ltd. must intend to address the information in question specifically (though not necessarily exclusively) to B Ltd., and that B Ltd. *must be aware* of having been specifically addressed. (This conscious giving and receiving of information is not to be confused with an agreement as to how the information should be used.)[194]

A more restrictive view has also been advocated, so that '[i]t is not sufficient for a competitor to disclose its intentions to another if this disclosure is not *sought or accepted* by the latter'.[195] However, in *Suiker Unie* the Court is clear that uncertainty about the future can be reduced if an undertaking

A. Ahlström Osakeyhtiö v Commission [1993] ECR I-1307, 1346, Case C-199/92 P *Hüls AG v Commission* [1999] ECR I-4287, para 162, Joined Cases T-25/95 and others *Cimenteries CBR v European Commission* [2000] ECR II-491, para 1389, and Wessely (2001: 745).

[190] Case T-1/89 *Rhône-Poulenc SA v Commission* [1991] ECR II-867, AG Opinion 941.

[191] Dashwood *et al.* (1976: 486). For critique of what evidences collusion: Mann (1973: 38–41) and Gijlstra and Murphy (1977: 59–60).

[192] Whish (2003: 489–496), Roth QC (2001: para 4.115–4.126), Capobianco (2004), and Kuhn (2001: 187–195).

[193] Motta (2004: 151–156), Van den Bergh and Camesasca (2001: 178–179), Neven *et al.* (1998: 66–70), Hylton (2003: 135), Case 48/69 *Imperial Chemical Industries v Commission* [1972] ECR 619, para 101–102, 112, and Whish (2003: 491).

[194] Dashwood *et al.* (1976: 486, emphasis added). Compare with Case CP/0871/01 *Price-Fixing of Replica Football Kit* [2003] 1 August, para 432, 436.

[195] Waelbroeck and Frignani (1999: 135, emphasis added). Also Joined cases C-89/85, C-104/85, C-114/85, C-116/85, C-117/85 and C-125/85 to C-129/85 *A. Ahlström Osakeyhtiö v Commission* [1993] ECR I-1307, AG Opinion para 169–174, Gijlstra and Murphy (1977: 59), Jones (1993: 275–276), van Gerven and Navarro Varona (1994: 599–601), and Wessely (2001: 757). Though the case ultimately did not turn on the issue, the claim is made in Case

'*disclose[s]*... the course of conduct which they themselves have decided to adopt or contemplate adopting on the market'.[196] All that would then seem to be required to evidence a concerted practice is the disclosure of information by a single market participant; there need be no commitment as to how such information will be used by the recipient. This latter position perhaps casts the net too wide, enabling undertakings to unwittingly become party to a concerted practice.[197] However, undertakings need not be willing parties to agreements.[198] Rather than deny there is agreement, the issue is treated as one going to responsibility and blame.[199] It is not obvious why the same approach should not be taken in relation to concerted practice.

1. *Jurisdictional nature of collusion*

It may be objected that a conception of concerted practice proven simply by evidence of communication, and a broad conception of communication, makes Article 81 EC too broad. However, the conception can be defended if the collusive terms enumerated in Article 81 EC are seen as jurisdictional, serving merely to capture conduct capable of causing the substantive harm the provision seeks to guard against. On this view, the view advanced here, since communication *may* be used to harm competition or 'inherently is fraught with anticompetitive risk' then its 'anticompetitive potential is sufficient to warrant scrutiny'.[200] Whether there is collusion and whether collusion has or will cause a restriction of competition in particular

CP/0001-02 *Collusive Tendering in Relation to Contracts for Flat Roofing Services in the West Midlands* [2004] 16 March, para 188–191, 197 and Case 1032/1/1/04 *Apex Asphalt and Paving Co Limited v Office of Fair Trading* [2005] CAT 4, para 142–143, 172–186. Compare with Case 1021/1/1/03 and 1022/1/1/03 *JJB Sports PLC v Office of Fair Trading; Allsports Limited v Office of Fair Trading* [2004] CAT 17, para 561–564, 571, 573–574, 583–586, 613, 616, 620–621, 652–655, 681, 698–703, 711.

[196] Joined Cases 40 to 48, 50, 54 to 56, 111, 113 and 114–73 *Coöperatieve Vereniging 'Suiker Unie' Ua v Commission* [1975] ECR 1663, para 174, emphasis added. Also Joined Cases T-202/98, T-204/98 and T-207/98 *Tate & Lyle PLC, British Sugar PLC, Napier Brown & Co. Ltd v Commission* [2001] ECR II-2035, para 35, 40, 54, Case 1021/1/1/03 and 1022/1/1/03 *JJB Sports PLC v Office of Fair Trading; Allsports Limited v Office of Fair Trading* [2004] CAT 17, para 158, and Faull and Nikpay (1999: para 2.28, 2.43).

[197] Case 1021/1/1/03 and 1022/1/1/03 *JJB Sports PLC v Office of Fair Trading; Allsports Limited v Office of Fair Trading* [2004] CAT 17, para 660, and Guerrin and Kyriazis (1993: 784).

[198] See discussion accompanying notes 34–38 above.

[199] Case C-453/99 *Courage Ltd v Bernard Crehan and Bernard Crehan v Courage Ltd and Others* [2001] ECR I-6297, para 34–36, AG Opinion para 40–44.

[200] Chief Justice Burger in *Copperweld Corporation v Independence Tube Corporation* 467 US 752, 768–769 (1944). Also Gellhorn *et al.* (2004: 265), Hildebrand (2002b: 26–27), and Ross (1993: 159).

circumstances are separate questions, though sometimes evidence of collusion overlaps with evidence that competition is restricted.[201]

In economics, there is nothing pejorative in the term collusion. For example, Stigler writes of merger as the most comprehensive form of *collusion*.[202] Firms collude when cooperation is more profitable than competition.[203] Collusion may be beneficial to both the undertakings and society.[204] Alternatively, collusion may be beneficial to the undertakings but detrimental to society.[205] Whish observes that in law collusion, tainted with pejorative meaning, is an 'opprobrious word' used to describe 'evil', and so seeks to avoid its use in circumstances not clearly characterized as such.[206] Though the existence of collusion is a question distinct from, and answered prior to, the question of whether collusion causes a restriction of competition, seeing collusion as pejorative at times leads to the two questions being fused.[207] Parties argue, and the Court sometimes accepts, that practices are not collusive if they do not impact upon competition.

One substantive issue of competitive impact sometimes considered as a jurisdictional question of whether there is collusion is the amount of uncertainty that would exist absent communication. A related factor often considered is the type of information communicated.[208] The Court has considered collusion absent if information communicated does not enhance certainty.[209] The essence of the argument is that the information conferred, received, or exchanged must be capable of harming competition in order to

[201] Guerrin and Kyriazis (1993: 797–802). The relevant harm is a substantive question considered in Chapter 5.

[202] Stigler (1983b: 41).

[203] Van den Bergh and Camesasca (2001: 170–171), Martin (2001: 50), and Lewis (1969: 5–24).

[204] Martin (2001: 65).

[205] Neven *et al.* (1998: 51–52) and Scherer and Ross (1990: 235–238). On the costs and benefits of collusion in various contexts: Cowen and Sutter (1999), Caplan and Stringham (2003) and Cowen and Sutter (2005).

[206] Whish (2003: 508–509).

[207] Gavil *et al.* (2002: 219–220) and Ross (1993: 162–171), note that the effect of certain collusive practices affects the assessment of whether or not collusion exists. This underlies the approach taken by Antunes (1991: 62), who considers that concerted practices require an anti-competitive purpose in order to be considered concerted practices.

[208] Waelbroeck and Frignani (1999: 138) and Guerrin and Kyriazis (1993: 787–788). On the substantive impact of the type of information exchanged: Capobianco (2004: 1256–1270), Kuhn (2001: 182–195), and Posner (1979b).

[209] Joined Cases C-89/85, C-104/85, C-114/85, C-116/85, C-117/85 and C-125/85 to C-129/85 *A. Ahlström Osakeyhtiö v Commission* [1993] ECR I-1307, para 64, Case 1021/1/1/03 and 1022/1/1/03 *JJB Sports PLC v Office of Fair Trading; Allsports Limited v Office of Fair Trading* [2004] CAT 17, para 657–662, and van Gerven and Navarro Varona (1994: 593–594). Compare with Case 1021/1/1/03 and 1022/1/1/03 *JJB Sports PLC v Office of Fair Trading; Allsports Limited v Office of Fair Trading* [2004] CAT 17, para 71, 368, 684–685, 835–838, 876 and Case CP/0871/01 *Price-Fixing of Replica Football Kit* [2003] 1 August, para 127–128, 131, 393, 416, 430.

be considered collusive. The operation of this consideration can be seen in *Tate & Lyle*. British Sugar and Tate & Lyle, competing sugar producers, held a meeting on 20 June 1986 at which British Sugar announced that it would end aggressive price competition.[210] During eight further meetings, retail prices were discussed and British Sugar gave its intended prices to Tate & Lyle.[211] It was argued that the information given and received did not reduce uncertainty to the extent necessary to restrict competition and so there was no collusion.[212] The Commission and CFI considered that uncertainty was reduced.[213] However, though the claim was rejected, it was considered—and considered as part of the issue of whether or not there was collusion.

Substantive issues are considered in the determination of communication in *Wood pulp*: customers informed suppliers of the prices charged by their competitors in order to negotiate a better deal.[214] Some pulp producers were aware of their competitors' prices as downstream customers.[215] The Commission argued that all that was required was that the information be reliable, regardless of how it was obtained.[216] The parties argued, and the Advocate General and Court accepted, that it was not communication to obtain information from customers, sales agents, and observation and analysis of the trade press.[217]

It is clear that information is obtained. It may be that certain methods of obtaining information are not communication.[218] Alternatively,

[210] Joined Cases T-202/98, T-204/98 and T-207/98 *Tate & Lyle PLC, British Sugar PLC, Napier Brown & Co. Ltd v Commission* [2001] ECR II-2035, para 9.

[211] Joined Cases T-202/98, T-204/98 and T-207/98 *Tate & Lyle PLC, British Sugar PLC, Napier Brown & Co. Ltd v Commission* [2001] ECR II-2035, para 11.

[212] Joined Cases T-202/98, T-204/98 and T-207/98 *Tate & Lyle PLC, British Sugar PLC, Napier Brown & Co. Ltd v Commission* [2001] ECR II-2035, para 34, 59. Compare Joined cases C-89/85, C-104/85, C-114/85, C-116/85, C-117/85 and C-125/85 to C-129/85 *A. Ahlström Osakeyhtiö v Commission* [1993] ECR I-1307, 1346.

[213] Joined Cases T-202/98, T-204/98 and T-207/98 *Tate & Lyle PLC, British Sugar PLC, Napier Brown & Co. Ltd v Commission* [2001] ECR II-2035, para 39, 60.

[214] Joined Cases C-89/85, C-104/85, C-114/85, C-116/85, C-117/85 and C-125/85 to C-129/85 *A. Ahlström Osakeyhtiö v Commission* [1993] ECR I-1307, 1345, 1367, Van den Bergh and Camesasca (2001: 179), and Stigler (1983b: 44).

[215] Joined Cases C-89/85, C-104/85, C-114/85, C-116/85, C-117/85 and C-125/85 to C-129/85 *A. Ahlström Osakeyhtiö v Commission* [1993] ECR I-1307, 1346, 1365, AG Opinion para 257.

[216] Joined Cases C-89/85, C-104/85, C-114/85, C-116/85, C-117/85 and C-125/85 to C-129/85 *A. Ahlström Osakeyhtiö v Commission* [1993] ECR I-1307, 1346–1347, AG Opinion para 246.

[217] Joined Cases C-89/85, C-104/85, C-114/85, C-116/85, C-117/85 and C-125/85 to C-129/85 *A. Ahlström Osakeyhtiö v Commission* [1993] ECR I-1307, 1365–1366, para 175–197, AG Opinion para 178–182, 247–278.

[218] See discussion at note 195 above.

jurisdiction to examine the conduct and its competitive impact is denied after a quasi-substantive assessment. It seems clear that, if the conduct was seen as communication, it does not cause a restriction on competition. There must be a causal connection between the competitive harm and collusion in order to establish a breach of Article 81(1) EC.[219] It is important to clarify the sense in which causation is used. To cause something is to intervene in the existing or expected state of the world in a way that makes a difference in the way things develop.[220] In order to determine causation the question to ask is whether the remaining set of conditions would produce the outcome in the absence of the condition at issue, ie, whether competition would be harmed absent the agreement, decision, or concerted practice under scrutiny.[221] The test is not satisfied by simply eliminating the scrutinized practice; it may be inappropriate to expect the parties to do nothing. Instead, the challenged practice must be replaced with a practice immune from challenge before asking if the same outcome would occur.[222] The sense one gains from reading the judgment is that no system could or ought to prevent customers seeking a better deal and in the process it is inevitable that they will reveal what deals are available elsewhere. The absence of causation, rather than the absence of collusion, saves the challenged conduct.

Another substantive issue of competitive impact sometimes considered as part of the jurisdictional assessment of whether there is collusion is the context in which the practice occurs. In identifying communication as the object of scrutiny within the conception of concerted practice the Court has indicated that it is because such conduct is capable of 'influenc[ing] the conduct on the market of an actual or potential *competitor*' that the Treaty 'strictly preclude[s] any direct or indirect contact *between such operators*'.[223] However, a concerted practice may be vertical as well as horizontal.[224] A practice occurring in a vertical context may have a different impact on

[219] Deringer (1968: 26–27). On the need for a causal link under the ECMR: Levy (2003: para 10.04(2), text accompanying notes 19–22).

[220] Hart and Honoré (1959: 27) and Honoré (1999: 2).

[221] Honoré (1999: 102–103). Hart and Honoré (1959: 30, 38–41, 69–72) consider that even when there are other conditions that can be described as the cause, voluntary human action intended to bring about the outcome is usually singled out.

[222] Honoré (1999: 103–107). It must be feasible for the parties to carry out the conduct used in the counterfactual hypothesis.

[223] Joined Cases 40 to 48, 50, 54 to 56, 111, 113 and 114–73 *Coöperatieve Vereniging 'Suiker Unie' Ua v Commission* [1975] ECR 1663, para 174, emphasis added.

[224] Goyder (2003: 77–78) and Waelbroeck and Frignani (1999: para 140).

competition than the same practice occurring in a horizontal context and at times the likely impact on competition seems to determine whether there is collusion. In *Tate & Lyle*, whilst British Sugar and Tate & Lyle processed sugar and offered it for sale, Napier Brown imported processed sugar and offered it for sale. The three firms were competitors. Napier Brown also purchased processed sugar from British Sugar and Tate & Lyle and was thus also in a vertical relationship with the producers.[225] At meetings held between 20 June 1986 and 13 June 1990, British Sugar gave Tate & Lyle and Napier Brown information about its future prices and production volumes.[226] Napier Brown argued that it was entitled to receive the information as a customer of British Sugar and so there was no collusion.[227] The CFI ruled that when participation can be characterized as horizontal or vertical, there is a rebuttable presumption that participation is horizontal.[228] There is also a duty on vertical participants not to facilitate or encourage communication between horizontal actors.[229] However, implicit in the argument and the ruling is that a concerted practice would not exist if participation was vertical.

It is clear that in a vertical relationship not all communication harms competition; it seems trite to state that it is legitimate for an undertaking to hold meetings with its customers.[230] Communication clearly occurs; the context in which information is exchanged, given, or received does not affect whether or not there is communication—context is relevant to determining the competitive impact of that communication. Apprehension

[225] Joined Cases T-202/98, T-204/98 and T-207/98 *Tate & Lyle PLC, British Sugar PLC, Napier Brown & Co. Ltd v Commission* [2001] ECR II-2035, para 5.

[226] Joined Cases T-202/98, T-204/98 and T-207/98 *Tate & Lyle PLC, British Sugar PLC, Napier Brown & Co. Ltd v Commission* [2001] ECR II-2035, para 10, 57.

[227] Joined Cases T-202/98, T-204/98 and T-207/98 *Tate & Lyle PLC, British Sugar PLC, Napier Brown & Co. Ltd v Commission* [2001] ECR II-2035, para 36.

[228] Joined Cases T-202/98, T-204/98 and T-207/98 *Tate & Lyle PLC, British Sugar PLC, Napier Brown & Co. Ltd v Commission* [2001] ECR II-2035, para 61–67.

[229] Joined Cases 100-103/80 *Musique Diffusion Française SA v Commission* [1983] ECR 1825, para 75–79, AG Opinion 1937–1938, Case 107/82 *AEG-Telefunken v Commission* [1983] ECR 3151, AG Opinion 3252, and Case CP/0871/01 *Price-Fixing of Replica Football Kit* [2003] 1 August, para 254–257, 317, 430. The existence and extent of this duty will be considered in Case T-36/05 *Coats Holdings Limited and J & P Coats Limited v Commission of the European Communities*.

[230] Joined Cases 100-103/80 *Musique Diffusion Française SA v Commission* [1983] ECR 1825, para 74, AG Opinion 1930, Case T-1/89 *Rhône-Poulenc SA v Commission* [1991] ECR II-867, AG Opinion 877, Case C-49/92 P *Commission v Anic Partecipazioni SpA* [1999] ECR I-4125, AG Opinion para 55–57, and Case C-199/92 P *Hüls AG v Commission* [1999] ECR I-4287, AG Opinion para 82–83.

of attaching what is viewed by some as the pejorative label of collusion prevents communication evidencing concerted practice, but at the same time denies proper substantive assessment by taking it beyond the scope of Article 81 EC. All communication should evidence concerted practice if the jurisdictional function of the collusive terms is accepted. In the same way that an agreement does not cease to be an agreement when it fails to cause competitive harm, concerted practices should not cease to be concerted practices when they do not harm competition. The jurisdictional and substantive questions are, and ought to remain, separate.

IV. Collusive outcomes without collusive conduct

Undertakings may achieve the outcome that would occur had they colluded, but without producing acceptable *evidence* of agreement, and without giving, receiving, or exchanging information: the parties act *as if* they colluded.[231] Undertakings are able to act *as if* they had agreed, or have given, received, or exchanged information if (a) a mutually beneficial strategy can be identified; (b) deviation from that strategy can be detected; (c) pressure can be brought to bear to prevent deviation from the mutually beneficial strategy.[232] A question dogging competition law is the extent to which acting *as if* there has been collusion falls within the boundary of Article 81 EC.[233]

In US antitrust, Rostow welcomed the possibility that such conduct could fall within the conception of collusion as a development that 'might produce not piddling changes in the detail of trade practice, but long strides towards the great social purpose of the statute'.[234] Inclusion of outcomes identical to those that would arise *if* undertakings had colluded, and ignoring the *cause* of those outcomes, would mean that the '[p]ainstaking search for scraps of evidence with a conspiratorial atmosphere are no longer necessary'.[235] Turner asks, '[i]f such a conclusion is repellent, why is it?'[236]

[231] Black (2003a: 68), Gavil *et al.* (2002: 245–246), and Hylton (2003: 140).

[232] Black (1992: 200–201), Lewis (1969: 52–68, particularly 60–68, on how knowledge that a mutually beneficial strategy is acquired in convention), Martin (2001: 50–53), Motta (2004: 140), Neven *et al.* (1998: 49–51), and Turner (1962: 659–663). The extent to which these conditions are ever satisfied in reality is considered in Posner (1969: 1566–1569).

[233] For the classic debate in US antitrust: Turner (1962) and Posner (1969), synthesized in Posner (2001: 55–60).

[234] Rostow (1947: 574). Also Hay (2000: 116).

[235] Rostow (1947: 585).

[236] Turner (1962: 663).

One objection is that the conduct is unilateral.[237] Replication of the outcome that would occur *if* undertakings had agreed, or given, received, or exchanged information:

is *not* an interaction back and forth between people. It is a process in which *one* [undertaking] works out the consequences of [its] beliefs about the world—a world [it] believes to include other [undertakings that] are working out the consequences of their beliefs.[238]

This objection is strong, but not compelling. In *Bayer*, conduct is considered unilateral when its aims can be achieved 'without the express or implied participation of another undertaking'.[239] However, the *as if* colluded outcome cannot be achieved if all undertakings do not opt for the same strategy—participation by others *is* necessary, even if only in a weak sense.[240]

A second objection to bringing conduct that produces the *as if* colluded outcome within the boundary of Article 81 EC is that such conduct cannot sensibly be said to cause a restriction on competition. Conduct is only said to cause an outcome if (a) the outcome would not occur absent the conduct and (b) there is conduct that can reasonably be expected to occur that does not cause the same outcome.[241] The undertakings have simply acted rationally; it would be irrational not to take competitors' reactions into account and absurd to require undertakings to act irrationally.[242]

The inability to capture outcomes that are the same as those that would be achieved had the undertakings colluded would leave a gap in the coverage of Article 81 EC. To the extent that a gap exists, the problem is structural and the market structure ought to be the focus of attack.[243] However, the size of the gap appears to be minimal. It will be recalled that to achieve the *as if* colluded outcome, undertakings must (a) identify a mutually beneficial strategy; (b) detect deviation from that strategy; and (c) prevent

[237] Hay (2000: 122–123), Neven *et al.* (1998: 51), Turner (1962: 665–666), and Van den Bergh and Camesasca (2001: 201).

[238] Lewis (1969: 32, emphasis in original).

[239] Case T-41/96 *Bayer AG v Commission* [2000] ECR II-3383, para 71. Also discussion in Section II.B.1(a)(ii), Black (2003c: 103) and Hylton (2003: 135–136, 140–143).

[240] Lewis (1969: 5–24), Posner (1969: 1576–1578), Turner (1962: 663–665), and Van den Bergh and Camesasca (2001: 171).

[241] See text accompanying notes 220–222 above. Compare Martin (2001: 63–64).

[242] Turner (1962: 669–672, 675–681).

[243] Hay (2000: 116–117, 123–124), Kaysen (1951: 270), Posner (1969: 1593–1605), Turner (1962: 665–672, 682–684), and Whish (2003: 507–508). Rostow (1947: 586–600) considers that it is an infringement of the competition rules that gives authority to impose a structural remedy but comments on the unwillingness of the authorities to impose such remedies.

deviation from the mutually beneficial strategy.[244] All of this must be achieved without producing acceptable *evidence* of agreement, and without giving, receiving, or exchanging information.[245] By considering these elements individually, it is shown that the gap through which *as if* colluded outcomes fall beyond the boundary of Article 81 EC is small.

A. *Identifying a mutually beneficial strategy*

Undertakings identify the mutually beneficial strategy:

> by putting [themselves] in the other fellow's shoes, to the best of [their] ability. If I know what you believe about the matters of fact that determine the likely effects of your alternative actions, and if I know that you possess a modicum of practical rationality, then I can replicate your practical reasoning to figure out what you will probably do, so that I can act appropriately.[246]

Transparency enables a mutually beneficial market strategy to be identified; each party can independently assess whether it is rational to act as if they have agreed to follow an anti-competitive strategy.[247] The identification of the mutually beneficial strategy does not involve communication.[248] So, in both *Dyestuffs* and *Sugar*, the Court confirmed that identifying the mutually beneficial strategy was not prohibited, writing that 'every producer is free to change his prices, taking into account in so doing the present or foreseeable conduct of his competitors', and that the concept of collusion 'does not deprive economic operators of the right to adapt themselves intelligently to the existing and anticipated conduct of their competitors'.[249]

B. *Monitoring adherence*

Deviation from the mutually beneficial strategy may occur for numerous reasons.[250] Particularly, deviation is profitable for an individual undertaking as long as other undertakings adhere to the strategy: 'the man who murders

[244] See references cited at note 232 above. [245] Posner (1969: 1569–1574).

[246] Lewis (1969: 27). [247] Black (2003b: 408) and Motta (2004: 150–151).

[248] Joined Cases C-89/85, C-104/85, C-114/85, C-116/85, C-117/85 and C-125/85 to C-129/85 *A. Ahlström Osakeyhtiö v Commission* [1993] ECR I-1307, 1365–1366, para 175–197, AG Opinion para 178–182, 247–278. Compare with Kaysen (1951: 264–269), considering that at the outset there is 'an agreement to agree'.

[249] Case 48/69 *Imperial Chemical Industries v Commission* [1972] ECR 619, para 118, and Joined Cases 40 to 48, 50, 54 to 56, 111, 113 and 114–73 *Coöperatieve Vereniging 'Suiker Unie' Ua v Commission* [1975] ECR 1663, para 174. [250] Posner (1969: 1570–1571).

the collusive price will receive the bequest of patronage'.[251] It must thus be possible to detect deviations from the mutually beneficial strategy.[252] The exchange of information facilitates monitoring of adherence to the mutually beneficial strategy.[253] An *agreement* to exchange information falls within the boundary of Article 81 EC.[254] It may be that no agreement exists, and that information exchange simply occurs by convention.[255] Lewis distinguishes convention from agreement: '[c]onvention turns out to be a general sense of common interest' that is 'mutually expressed and is known to both'. However, unlike agreement conformity does not rely on 'the interposition of promise'.[256] Convention and agreement are distinct because the former does not give rise to obligation. As obligation is lacking, convention does not create complete certainty as to how others will act.[257] Two reasons exist for convention being preferred over other methods of uncertainty-reduction: first, other means of coordination may be prohibited; second, convention may provide a satisfactory level of certainty.[258]

If we acknowledge that 'a convention that it would be in each person's interest to observe if everyone else observed it will be established and maintained without any special mechanism of commitment or enforcement', the question must be whether conventions come within the scope of collusion.[259] It seems that they must, particularly since convention can be created by agreement—though the 'direct influence fades away in days, years, or lifetimes . . . the indirect influence of the agreement is constantly renewed, and in time it comes to predominate. . . . a convention created by agreement is no longer different from one created otherwise: it bears no trace of its origin.'[260] Absent an agreement to exchange, the giving, receiving, or exchange of information remains communication establishing

[251] Stigler (1983b: 44). Also Scherer and Ross (1990: 237, 244–248, 308–311), Carlton and Perloff (2000: 121–152), and Phlips (1995: 49–51).

[252] Stigler (1983b: 42–56), Carlton and Perloff (2000: 125), and Van den Bergh and Camesasca (2001: 174–182).

[253] Kuhn (2001: 187–195) and Posner (2001: 86–87).

[254] Whish (2003: 489–496), Roth QC (2001: para 4-115–4-126), Capobianco (2004), and Kuhn (2001: 187–195).

[255] Joined Cases C-89/85, C-104/85, C-114/85, C-116/85, C-117/85 and C-125/85 to C-129/85 *A. Ahlström Osakeyhtiö v Commission* [1993] ECR I-1307, 1367–1368, para 77–78.

[256] Lewis (1969: 3–4).

[257] Lewis (1969: 24–36, and on the level of coordination that can be achieved by convention 36–82).

[258] Lewis (1969: 35, 46–47).

[259] Fried (1981: 15).

[260] Lewis (1969: 84, also 83–88), Case 51/75 *EMI Records v CBS United Kingdom Limited* [1976] ECR 811, para 30–31, Faull and Nikpay (1999: para 2.30), Martin (2001: 64), Neven *et al.* (1998: 62–63), and Van den Bergh and Camesasca (2001: 176, 181–182).

the existence of a concerted practice regardless of the reason for the communication.[261] Such communication makes the market transparent, and is rightly subject to scrutiny.[262] It should be recalled that communication is not prohibited. Instead, communication gives rise to the jurisdiction to examine the competitive consequences of the practice.[263] If there is a gap which the concepts of collusion fail to cover, it is only when the natural transparency of the market enables firms to monitor adherence to the mutually beneficial strategy.[264]

C. *Deterring deviation*

If undertakings identify a mutually beneficial strategy and compliance with that strategy can be monitored, there is no incentive to deviate from the strategy. Additional sales cannot be won by offering discounts, as these will be seen and matched by others: this of course assumes that there is no delay between deviation and detection. In addition, since deviation involves an increase in output, it is also assumed that other undertakings have the capacity to increase supply in response to deviation.[265] If these preconditions are met, deviation simply results in lower priced sales but not an increase in sales: there is no incentive to deviate from the *as if* colluded outcome.[266] This confirms that whilst acting *as if* there has been collusion is not prohibited, scrutiny should be placed on the mechanisms by which deviation is monitored.

V. Conclusion

The aim of this chapter has been to identify and develop the concepts of agreement and concerted practice used in Article 81 EC. It has argued that both agreement and concerted practice are methods by which undertakings can reduce uncertainty about the future conduct of others. In this sense, agreement and concerted practice have 'the same nature and are only distinguishable from each other by their intensity and the forms in which

[261] Case 48/69 *Imperial Chemical Industries v Commission* [1972] ECR 619, para 101–102, 112, Hylton (2003: 135), Motta (2004: 151–156), Neven *et al.* (1998: 66–70), Van den Bergh and Camesasca (2001: 178–179), and Whish (2003: 491), suggest an affirmative answer.

[262] Phlips (1995: 81–93) and Hay (2000).

[263] Hay (2000: 124–125).

[264] Hay (2000: 125–127), Kuhn (2001: 172), and Martin (2001: 50–57).

[265] Posner (2001: 57–59).

[266] Phlips (1995: 52–54, 94–105) and Kuhn (2001: 172).

they manifest themselves'.[267] This also makes it unnecessary to distinguish between them at the stage of substantive assessment.[268] However, the evidence required to establish that a particular form of collusion exists does differ.[269] Agreement requires common intention. This can be shown to exist when there is evidence of communication (offer) and commitment (acceptance). A concerted practice does not require commitment and evidence of communication is sufficient. What constitutes communication remains uncertain.

The conception of collusion capturing means by which uncertainty as to the future conduct of others is reduced identifies conduct appropriate for scrutiny under Article 81 EC as it is uncertainty as to the future that forces undertakings to engage in a competitive struggle. Reduced uncertainty provides comfort for those who 'prefer . . . the security of keeping what they already ha[ve] to the chance of winning more, when this entail[s] the risk of total loss'.[270] The collusive terms should be seen as serving a jurisdictional function; this makes it possible to understand and tolerate the broad reach. The function played by the concepts of agreement, decision and concerted practice, like that of undertaking explained in Chapter 3, is to bring within the boundary of Article 81 EC conduct that is appropriately subject to the substantive assessment explained in Chapters 5 and 6. Finding collusive conduct between undertakings still leaves the task of substantive assessment.

[267] Case C-49/92 P *Commission v Anic Partecipazioni SpA* [1999] ECR I-4125, para 131.

[268] *Argos Limited and Littlewoods Limited v Office of Fair Trading* [2004] CAT 24, para 665, Case 48/69 *Imperial Chemical Industries v Commission* [1972] ECR 619, para 64, Case C-49/92 P *Commission v Anic Partecipazioni SpA* [1999] ECR I-4125, para 108, Case T-1/89 *Rhône-Poulenc SA v Commission* [1991] ECR II-867, AG Opinion 944, Antunes (1991: 71–77), Deringer (1968: 10), Faull and Nikpay (1999: para 2.54–2.55), Jakobsen and Broberg (2002: 127), and Joshua and Jordan (2004). Black (2003c: 107–108) states that 'the statement X and Y collude means that X and Y have either an agreement or a concerted practice'.

[269] Antunes (1991: 66), Dashwood *et al.* (1976: 481), Gijlstra and Murphy (1977: 58), and Steindorff (1972: 509).

[270] Deringer (1962: 1). Also Guerrin and Kyriazis (1993: 774).

5

The meaning and existence of restricted competition

I. Introduction

In the summer of 1997, Whish expressed the view that 'the debate about what is meant by a restriction of competition under Art. [81(1)] has been with us for 30 years, but I do not believe we are any closer to an acceptable solution to this central conundrum of competition law'.[1] This chapter is concerned with the central conundrum of what is meant by a restriction of competition and attempts to bring us closer to an acceptable solution. An acceptable solution has eluded us because consensus on the function the term 'restriction of competition' is supposed to serve within Article 81(1) EC is lacking. Goyder draws attention to how the function the requirement of restricting competition is supposed to serve determines its meaning, writing that:

> [t]oo strict a definition might cover almost every agreement . . . Too liberal a definition would reduce the jurisdiction of the Commission . . . In determining the correct approach to these words within Article 81(1) it is necessary to decide whether the Article is primarily designed to *assert jurisdiction*, or whether it seeks to *provide an assessment* of whether individual agreements are justified.[2]

Section II considers the view that restriction of competition serves a jurisdictional function. Jurisdiction is claimed over all practices with the potential to impede Community integration; the substantive assessment is carried out under Article 81(3) EC. Section III considers the view that restriction of competition in Article 81(1) EC serves a substantive function. The challenge is to identify a substantive obligation appropriate to impose

[1] Panel Discussion (1998: 461). [2] Goyder (2003: 91), emphasis added.

on undertakings engaged in collusion. Since undertakings are engaged in economic activity, the substantive obligation must relate to economic activity. It is argued that the substantive obligation placed on those engaged in economic activity is not to collude to cause, or engage in collusion causing, economic inefficiency. Whilst there are two aspects of efficiency (allocative and productive/dynamic), Article 81(1) EC is purely an allocative efficiency enquiry.[3] This meaning is shown by examining the factors causing the Court to proclaim that restricted competition exists. Examination reveals that collusion is said to have the *effect* of restricting competition when allocative inefficiency is measured or predicted; collusion is said to have the *object* of restricting competition when a legal presumption of allocative inefficiency exists. Since a restriction of competition can always and only be said to exist when collusion causes allocative inefficiency, restriction of competition as a substantive element in Article 81(1) EC and allocative inefficiency are synonymous. Section IV draws conclusions from this enquiry and makes clear how the object and the effect criteria operate in Article 81(1) EC.

II. Restriction of competition as a jurisdictional element

The restriction of competition element in Article 81 EC can serve as a jurisdictional rather than substantive requirement.[4] Rather than call for a substantive assessment, a restriction of competition within the meaning of Article 81(1) EC exists, and thus Community competition law asserts jurisdiction, when the conduct under scrutiny is 'inimical to the objectives of the Community'.[5] Sometimes jurisdiction is asserted whenever a purchaser loses an element of the freedom to determine what to do with the goods or services procured.[6] This reflects the fact that competition law is valued by some for its ability to protect and promote certain freedoms.[7] At other times, and the main focus here, jurisdiction is asserted over all conduct with the ability to 'perpetuate the compartmentalization of the

[3] It is argued in Chapter 6 that productive and dynamic efficiency issues are considered within Article 81(3) EC.

[4] Verouden (2003: 532–533), Goyder (2003: 91), Amato (1997: 117–118), Collins (1984: 522), Forrester and Norall (1984: 13, 22), and Neven *et al.* (1998: 37).

[5] Whish (2001: 95).

[6] Constance (1999: 482), Wesseling (2000: 27, 31, 82–83, 88–93), Petersmann (1999: 151–154), Siragusa (1998: 653–654), Hawk (1988: 69–70), and Hawk (1995: 973, 978–979).

[7] Chapter 2, Section IV

Community'.[8] Such conduct is deemed to warrant substantive assessment under Article 81(3) EC 'as a matter of policy'.[9] This reflects the fact that competition law is valued by some for its ability to promote market integration.[10]

The use of 'restriction of competition' as a jurisdictional element and of market integration as the basis on which jurisdiction is claimed has at its origin *Consten & Grundig*, it being considered that:

an agreement . . . which might tend to restore the national divisions in trade between Member States might be such as to frustrate the most fundamental [objectives] of the Community. The Treaty, whose preamble and content aim at abolishing the barriers between states, and which in several provisions gives evidence of a stern attitude with regard to their reappearance, could not allow undertakings to reconstruct such barriers. Article [81(1)] is designed to pursue this aim.[11]

Seeing market division as antithetical to a single market, the Commission considers that 'Article [81(1)] applies *virtually automatically* to . . . agreements which establish absolute territorial protection for exclusive distributors. This point is central to Commission policy.'[12] Whish notes that '[u]nification of the single market is an obsession of the Community authorities . . . Faced with a conflict between the narrow interests of a particular firm and the broader problem of integrating the market, the tendency will be to subordinate the former for the latter.'[13] Korah goes on to write that the Commission 'saw exclusive distribution agreements as dividing the common market . . . So, the Commission considered that most exclusive agreements infringed article 81(1).'[14] The conclusion drawn is that there can be no surprise that 'a system of competition policy which does not pretend to be implemented on the basis of economic analysis should reveal itself to have economically indefensible consequences'.[15] Thus, in

[8] Green Paper on Vertical Restraints in EC Competition Policy Com (96) 721 Final [1997] 4 CMLR 519, para 117. Also Boscheck (2000: 24–25), Cosma and Whish (2003: 41–42), Faull and Nikpay (1999: para 6.26, 7.07–7.10, 7.95–7.100), Hays (2001), Korah and O'Sullivan (2002: 21–23), Mastromanolis (1995: 561–565, 573–588, 607–608), Siragusa (1998: 656–658), Verouden (2003: 530), and Wesseling (2000: 80–82, 85–88).

[9] Whish (2001: 95).

[10] Chapter 2, Section III

[11] Cases 56 and 58–64 *Établissements Consten Sàrl and Grundig-Verkaufs-GmbH v Commission* [1966] ECR 299, 340.

[12] Green Paper on Vertical Restraints in EC Competition Policy Com (96) 721 Final [1997] 4 CMLR 519, executive summary para 21, italics added, underlining in original. Also Korah and O'Sullivan (2002: 104–108).

[13] Whish (2001: 19).

[14] Korah (2000: 237). Also Neven *et al.* (1998: 37–38).

[15] Neven *et al.* (1998: 44).

DRU/Blondel, the Commission considered that the restriction of competition element was satisfied even after finding that the collusion *did not* lead 'to a sales price higher than in its country of origin' and that not only had the undertaking 'lowered the sales price by reducing his margin of profit', but also could not increase the price 'because of the threat of importation by other undertakings purchasing the products in question from a middleman outside the concession area'.[16]

The jurisdictional role assigned to the 'restriction of competition' element in Article 81(1) EC can be seen in the *de minimis* doctrine. In *Franz Völk v Vervaecke*, considering that collusive conduct concerned only between 0.2 per cent and 0.05 per cent of total production in Germany, the Court ruled that 'an agreement falls outside the prohibition in Article [81] when it has only an insignificant effect on the markets, taking into account the weak position which the persons concerned have on the market of the product in question'.[17] This can be read as involving a substantive assessment.[18] However, the doctrine may also be seen to serve a jurisdictional function. The Commission's 2001 notice on agreements of minor importance makes it clear that failure to meet the non-appreciability criteria *does not* show that an applicable restriction of competition exists.[19] It is also clear that the non-appreciability criteria can be satisfied by collusion that *does* appreciably restrict competition.[20] Thus, no substantive claim that a restriction of competition does or does not exist is ever made. Instead it seems that the *de minimis* doctrine is used as a jurisdictional tool through which the Commission proclaims that it will not assert jurisdiction over horizontal or vertical agreements between parties each with a market share of below 10 per cent and 15 per cent respectively, but will always assert jurisdiction over what it terms hardcore restrictions.[21]

[16] *Dru/Blondel* [1965] JO 2194/65 [1965] CMLR 180, 183. Also *Hummel/Isbecque* [1965] JO 2581/65 [1965] CMLR 242 and *Maison Jallatte* [1966] JO 37/66 [1966] CMLR D1.

[17] Case 5/69 *Franz Völk v Sprl Ets J Vervaecke* [1969] ECR 295, para 5/7. Also Case 1/71 *Société Anonyme Cadillon v Firma Höss* [1971] ECR 351, para 9 and Goyder (2003: 84–85).

[18] Whish (2003: 132) and Communication from the Commission on the Application of the Community Competition Rules to Vertical Restraints (Follow-up to the Green Paper on Vertical Restraints) Com(98)544 Final [1998] OJ C 365/3, 21.

[19] Commission Notice on Agreements of Minor Importance (de Minimis) [2001] OJ C 368/13, para 2.

[20] Commission Notice on Agreements of Minor Importance (de Minimis) [2001] OJ C 368/13, para 8.

[21] Commission Notice on Agreements of Minor Importance (de Minimis) [2001] OJ C 368/13, para 4, 7(b), 11 and Verouden (2003: 548–550). Compare with the account given in Waelbroeck and Frignani (1999: 173).

Viewing restriction of competition in Article 81(1) EC as a jurisdictional requirement leaves all substantive assessment for Article 81(3) EC.[22] Assigning a jurisdictional role to the restriction of competition element did not merely determine that Community competition law applied. It also determined the institution that would carry out the substantive assessment; to ensure uniform application in a Community that did not have a culture of competition, Regulation 17/62 conferred on the Commission sole competence to carry out the substantive assessment under Article 81(3) EC.[23] The jurisdictional interpretation of restriction of competition 'exclud[ed] national courts from the more important part of the antitrust analysis'.[24]

The jurisdictional interpretation resulted in complaints that 'the Commission too broadly applies Article [81(1)] to agreements having little or no anticompetitive effects'.[25] However, on the jurisdictional interpretation the purpose of Article 81(1) EC was never to provide a substantive assessment. The jurisdictional approach to the term 'restriction of competition' has probably been abandoned, but is definitely on the wane, for two reasons. First, the formal completion of the internal market diminishes the extent to which integration as an ideological constituent of the conception of competition can be supported.[26] Second, Regulation 1/2003 now fully decentralizes the application of Article 81 EC.[27] It is now not only the Commission that applies Article 81(3) EC, but also national courts and competition authorities. Thus, an interpretation of Article 81(1) EC can no longer be used to allocate competence.[28]

[22] Goyder (2003: 92). Compare with Gyselen (2002: 189).

[23] Article 9(1) of Regulation 17. First Regulation Implementing Articles 81 and 82 of the Treaty [1962] OJ Special Edition 204/62, White Paper on Modernisation of the Rules Implementing Articles 85 and 86 of the EC Treaty Commission Programme No 99/027 [1999] OJ C 132/1, para 4, Fox (2001: 142), and Schaub (1999: 154).

[24] Hawk (1988: 65). Also Manzini (2002: 394).

[25] Hawk (1995: 974). Also Green Paper on Vertical Restraints in EC Competition Policy Com (96) 721 Final [1997] 4 CMLR 519, executive summary para 37, Wesseling (2000: 88–93), Hawk (1995: 977–978, 982–983), Hawk (1988: 64–65, 69–70), Faull and Nikpay (1999: para 2.57, 7.40), Green (1988: 199–201), Korah (2000: 237), Korah and Rothnie (1992: 243), Nazerali and Cowan (2000: 51), Whish (1993: 564), Siragusa (1998: 674–678), Bright (1995: 515), Forrester and Norall (1984: 23–24, 38–40), and Hildebrand (1998: 188, 219, 235).

[26] Gerber (1994: 143–145), Wesseling (2000: 48–50), and Chapter 2, text accompanying notes 60–80.

[27] Article 1 of Council Regulation (EC) No 1/2003 [2003] OJ L 1/1. On the new jurisdictional settlement: Lenaerts and Gerard (2004: 319–328), Paulweber (2000), and Ehlermann (2000).

[28] Bishop (2003: 231–232).

III. Restriction of competition as a substantive element

A substantive meaning can be ascribed to the term 'restriction of competition' in Article 81(1) EC. The challenge is to identify a substantive obligation appropriate to impose on entities engaged in economic activity (undertakings). Principles of interpretation would seem to demand that since the obligation is only imposed on those engaged in economic activity the obligation must relate to economic activity, which consists of offering goods or services to the market place.[29] It can be shown that, as a substantive matter, the term restriction of competition in Article 81(1) EC is concerned with contrived scarcity of output—the substantive obligation placed on those engaged in economic activity is not to collude to cause allocative inefficiency.[30]

Article 81(1) EC is concerned with the allocative inefficiency problem discussed in Chapter 2.[31] The problem is represented diagrammatically as in Figure 5.1.[32] The downward sloping curve represents market demand. The demand curve slopes downwards because the cheaper the price, the greater the number of consumers willing and able to buy the goods. P_c and Q_c show the price and output at which society is best off. However, undertakings can advantage themselves relative to others by contriving a scarcity of output from Q_c to Q_a, which results in a price increase from P_c to P_a. When scarcity is contrived the demand of consumers willing to pay more than the cost of production, but less than the price actually charged, goes unsatisfied. The consequence of contrived scarcity is that society has insufficient desirable goods and pays too much for them.[33] This allocative inefficiency consequence, termed dead-weight loss, is labelled as area A: it is with area A that Article 81(1) EC is concerned.

That the substantive meaning of restriction of competition within Article 81(1) EC is allocative inefficiency is shown by examining the factors causing the Court to proclaim that restricted competition exists.

[29] Fuller (1969: 81–91).

[30] Compare with Guidelines on Vertical Restraints [2000] OJ C 291/1, para 100–102, Jenny (1994: 197–212), and Collins 1984: 506–508).

[31] Chapter 2, Section II.A.

[32] Williamson (1968a). Bork (1993: 108) considers the basic theory generally applicable to all competition law problems.

[33] In Cases 1035/1/1/04 and 1041/2/1/04 *The Racecourse Association and Others v Office of Fair Trading* [2005] CAT 29, para 178–179, 183, 189, 191, 202, the UK CAT considered that a restriction of competition involved collusion causing higher prices.

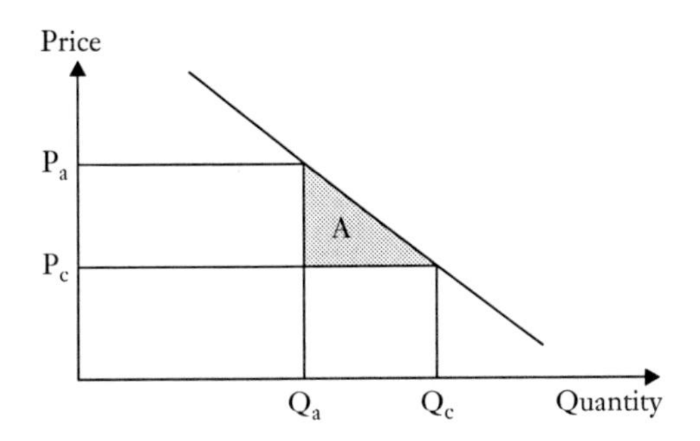

Figure 5.1 Article 81(1) EC as an allocative inefficiency enquiry.

Article 81(1) EC allows allocative inefficiency to be shown in two ways: an agreement may be said to cause allocative inefficiency by its object or by its effect.[34] Examination reveals that collusion is said to have the *effect* of restricting competition when allocative inefficiency is measured or predicted; collusion is said to have the *object* of restricting competition when a legal presumption of allocative inefficiency exists. Since a restriction of competition can always be said to exist when collusion is shown to cause allocative inefficiency, restriction of competition as a substantive element in Article 81(1) EC and allocative inefficiency are synonymous. The methods by which allocative inefficiency is shown to exist—measurement, prediction, and presumption—are considered below.

A. *Measurement*

Competition is said to be restricted within the meaning of Article 81(1) EC, and allocative inefficiency shown to exist, when we know that output has fallen or prices have risen as a consequence of the collusion under scrutiny.[35] This requires *propositional knowledge*, traditionally defined as justified true belief.[36] In this form, we *know* that output has fallen and price has risen, if (1) output has fallen and price has risen, (2) we believe that output has fallen and price has risen, and (3) we have adequate justification for believing that output has fallen and price has risen. The first condition is called the truth condition and a proposition is true if it corresponds to a

[34] This entails a rejection of the 'distinctness thesis' but acceptance of the 'evidence thesis' as articulated in Black (2005: Chapter 4).

[35] The causal element will usually be assumed in the remainder of the discussion.

[36] Everitt and Fisher (1995: 17) and Sturgeon (1995). These conditions are regarded as necessary, but not always sufficient.

fact.[37] The fact of reduced output, increased price, and the causal link between this and the collusion under scrutiny can be ascertained by measuring market output and price before and after implementation of an agreement said to be the cause of the change.[38] Thus Evans writes '[t]he only trustworthy way of finding out whether business practices harm consumers is to examine their impact on consumers. *Have they raised prices, restricted output, or reduced quality*?'[39] This takes what Salop calls the 'first principles' approach to a competition law enquiry.[40]

The idea of observing reduced market output and increased market price has much appeal. In *ICI v Commission*, allocative inefficiency was directly observed; the price of aniline dyes increased by 15 per cent between 7 and 20 January 1964, 10 per cent on 1 January 1965, and between 8 per cent and 12 per cent in October 1967.[41] The contested issue in the case was not the reduction in output/increase in price, but whether the cause was collusion.[42] In *Miller International* competition was said to be restricted because collusion resulted in increased prices; the agreement enabled Miller to charge 'its German customers prices differing sharply from the export prices, the latter being lower than the prices charged to wholesalers and much lower than the prices of products supplied to department stores, retail trade organizations, retailers and private consumers'.[43] In *Konica*, competition was said to be restricted by an agreement that enabled the undertaking to charge 30 per cent more in Germany than was charged in the UK.[44] In *Zera/Montedison* competition was found to be restricted by

[37] The classic formulation of the correspondence theory of truth is given by Russell (1980: 75). Compare with Sainsbury (1995: 105–114).

[38] Case 56/65 *Société Technique Minière v Maschinenbau Ulm GmbH* [1966] ECR 235, 250, Guidelines on the Application of Article 81(3) of the Treaty [2004] OJ C 101/97, para 17, Easterbrook (1984: 31–33), Posner (1977: 18–19), Posner (1981b: 21), Joffe (2001: text accompanying notes 15 and 31), and Nealis (2000: 359).

[39] Evans (2001: 545–546, emphasis added).

[40] Salop (2000).

[41] Case 48/69 *Imperial Chemical Industries v Commission* [1972] ECR 619, para 1–6, 83–87.

[42] There was some dispute at para 58 as to the correct measure of price.

[43] Case 19/77 *Miller International Schallplatten GmbH v Commission* [1978] ECR 131, para 5.

[44] *Konica* [1988] OJ L 78/34 [1988] 4 CMLR 848. Also *Preserved Mushrooms* [1975] OJ L 29/26 [1975] 1 CMLR D83, IFTRA [1975] OJ L 228/3 [1975] 2 CMLR D20, *Bomée-Stichting* [1975] OJ L 329/30 [1975] 1 CMLR D1, *Pabst and Richarz/BNIA* [1976] OJ L 331/24 [1976] 2 CMLR D63, Reuter/BASF [1976] OJ L 254/40 [1976] 2 CMLR D44, *Centraal Bureau Voor De Rijwielhandel* [1978] OJ L 20/18 [1978] 2 CMLR 194, *Videocassette Recorders* [1978] OJ L 47/42 [1978] 2 CMLR 160, *Pioneer* [1980] OJ L 60/21 [1980] 1 CMLR 457, *Hasselblad* [1982] OJ L 161/18 [1982] 2 CMLR 233, *Windsurfing* International [1983] OJ L 229/1 [1984] 1 CMLR 1, *Miller International Schallplatten GmbH* [1976] OJ L 357/40 [1977] 1 CMLR D61, *Theal/Watts* [1977] OJ L 39/19 [1977] 1 CMLR D44, *Ford Agricultural* [1993] OJ L 20/1 [1995] 5 CMLR 89, and *Advocaat Zwarte Kip* [1974] OJ L 237/12 [1974] 2 CMLR D79.

collusion enabling an undertaking to charge 20 per cent more in Germany than was charged in Italy.[45] In *Tepea v Commission* competition was said to be restricted by collusion that enabled an undertaking to charge 32 per cent more in the Netherlands than was charged in the UK.[46]

For some, there is simply no better way of confirming whether price and output have changed than by looking at price and output.[47] For others, direct measurement is ruled out as too unwieldy, complex and time consuming.[48] One limitation is that the measurement of price and output can only occur *ex post*. This makes the approach unsuitable for what has traditionally been a major task in the Community enforcement of Article 81(1) EC—*ex ante* control.[49] However, from 1 May 2004 there was a shift from *ex ante* to a system of *ex post* control.[50] The switch to *ex post* control means that direct measurement will play an increasing role in competition law litigation and enforcement so there is a need to clarify the technique. One issue requiring clarification is the extent to which market definition is necessary and how it is to be carried out using the direct measurement approach. In *Consten and Grundig* competition could be said to be restricted by collusion since it resulted in prices 20 per cent higher in France than in Germany.[51] Whilst we may observe that a *firm's* price is higher or its output lower than would be the case absent collusion, the allocative inefficiency problem arises with changes in *market* output and price.[52] Though Grundig's price

[45] *Zera/Montedison and Hinkens/Stähler* [1993] OJ L272/28 [1995] 5 CMLR 320, para 22–32, 52.

[46] Case 28/77 *Tepea BV v Commission* [1978] ECR 1391, para 56.

[47] *Appalachian Coals v United States* 288 US 344, 377–378 (1933), Areeda (1986: 577), Coate and Fischer (2001: 796, 805–806), Gavil (2000a: 89, 91–92 text accompanying note 20, 95–102), Patterson (2000: 432 text accompanying note 21), American Bar Association (Antitrust Section) and Hartley (chairman) (1999: 107–108), Carrier (1999: 1329–1334), and Gellhorn and Tatham (1984/1985: 170–172). Compare with Coppi and Dobson (2002).

[48] Adams (1991), Asch (1970, 1984 reprint: 111–114, 165, 180–182), Bork (1993: 117, 123–129), Kahn (1953: 37–41), Markovits (1975a: 846–847), Nicolaides (2000: 9), Posner (1976: 22–25), and Viscusi *et al.* (2000: 86–88).

[49] For strong support for the focus on *ex ante* control see Monopolkommission (2000: 16–19, 27, 29). For the origin of this position see Deringer (1968: para 401, 410–413) and Deringer (1962: 4–8). On the conditions under which *ex ante* enforcement detects and deters more anticompetitive agreements than *ex post* enforcement, and vice versa, see Hahn (2000), Van den Bergh and Camesasca (2001: 144–145), and Nealis (2000: 360).

[50] Council Regulation (EC) No 1/2003 [2003] OJ L 1/1 and Schaub (1999: 132–136). For the background: White Paper on Modernisation of the Rules Implementing Articles 85 and 86 of the EC Treaty Commission Programme No 99/027 [1999] OJ C 132/1, Paulweber (2000), and Ehlermann (2000).

[51] *Grundig/Consten* [1964] JO 2545/64 [1964] CMLR 489, 493, Cases 56 and 58–64 *Établissements Consten Sàrl and Grundig-Verkaufs-GmbH v Commission* [1966] ECR 299, 343, and Korah (1986: 93).

[52] Klein (1993: 44, 71–85) and Verouden (2003: 528).

increased, we know nothing of the impact this had on market price, which may have reduced.[53] The consequence for market price is revealed by weighing the reduced output of a particular undertaking against the increased output of other undertakings attracted to the market; market definition would then seem to be required.

Markets are determined by consumer perceptions of uniqueness.[54] It may be that the boundaries of the market can simply be observed. In *Grundig/Consten*, the Commission considered that 'for *trade mark* articles . . . the products of various manufacturers present *different* external, and in part also technical, characteristics'.[55] The Court endorsed this approach on the basis that 'the more producers succeed in their efforts to render *their own makes of product* individually distinct in the eyes of the consumer, the more the effectiveness of competition between producers tends to diminish'.[56] Brands are certainly economically significant.[57] Moreover, the use of brands to define markets has received some treatment in the economic literature.[58] The importance of brands in the Commission's assessment of market conduct has been recognised.[59] In *DRU/Blondel*, the Commission considered that 'the products of different manufacturers are distinguished by their individual characteristics; this individual characterisation, which is *emphasised by trade marks*, allows the consumers to show preferences'.[60] By way of contrast, in *Davidson Rubber*, the fact that the manufacturers '*use no trade marks* for these articles' prevented the treatment of the firm as the market.[61] In *Volkswagen* the Court makes it clear that market definition is only required when there is no other way to show that competition is restricted within the meaning of Article 81(1) EC.[62] Moreover, it is clear that at times branding is used as the hallmark of a consumer perception of

[53] Cases 56 and 58–64 *Établissements Consten Sàrl and Grundig-Verkaufs-GmbH v Commission* [1966] ECR 299, 342, Korah (2000: 60–61, 237), Korah and Rothnie (1992: 37–38, 298–299), Korah (1986: 93), and Korah (1984: 6–9, 31–32).

[54] Ferry (1981: 213, 216), Areeda and Kaplow (1997: 19–24), Maurice and Phillips (1986: 477–485), Carlton and Perloff (2000: 196–215), Asch (1970, 1984 reprint: 159, 177).

[55] *Grundig/Consten* [1964] JO 2545/64 [1964] CMLR 489, 496, emphasis added.

[56] Cases 56 and 58–64 *Établissements Consten Sàrl and Grundig-Verkaufs-GmbH v Commission* [1966] ECR 299, 343. Also Case 161/84 *Pronuptia de Paris GmbH v Pronuptia de Paris Irmgard Schillgalis* [1986] ECR 353, para 24.

[57] Chamberlin (1950) and Klein (2001).

[58] Grimes (1995: 95), Schmalensee (1982b), Werden (1998: 398–409), and Schmalensee (1983). On the relevance of branding to market definition in merger control: Levy (2003: para 8.04(2)(a)).

[59] Verouden (2003: 547, 549) and Korah (2000: 60).

[60] *Dru/Blondel* [1965] JO 2194/65 [1965] CMLR 180, 181, emphasis added.

[61] *Davidson Rubber Co* [1972] JO L 143/31 [1972] CMLR D52, para 28, emphasis added.

[62] Case T-62/98 *Volkswagen AG v Commission* [2000] ECR II-2707, para 230–231. Also Keyte and Stoll (2004).

uniqueness, which determines the boundaries of the market.[63] It remains to be seen whether and when a more sophisticated approach to market definition is required.[64]

B. *Prediction*

Article 81(1) EC is not only concerned with collusion that *has* restricted competition but also with preventing restrictions on competition occurring.[65] As Lutz notes, '[t]he main usefulness of anti-trust policies lies in their deterrent effect; and anyone who knows the situation in America knows that this effect is extremely powerful.'[66] Deterrence is enhanced by the power to sanction infringements.[67] Since *ex ante* intervention prevents reduced output and increased prices from occurring, we cannot have knowledge of allocative inefficiency.[68] In the absence of knowledge, the standard we aspire to is justified belief.[69] Instead of requiring output to have fallen and prices to have risen, good grounds for believing that, absent intervention, output *will* fall or prices *will* rise as a consequence of collusion are sufficient for intervention. The standard of justified belief is acceptable because of the relationship between such a belief, evidence, and truth. The greater the evidence that something is true, the more justified we are in our belief that it is true, and vice versa.[70]

A restriction of competition within the meaning of Article 81(1) EC covers both actual and potential allocative inefficiency.[71] Whilst actual

[63] Other decisions using this observational technique of market definition are *Goodyear Italiana/Euram* [1975] OJ L 38/10 [1975] 1 CMLR D31, involving 'Vitafilm' food wrap; *Duro-Dyne/Europair* [1975] OJ L 29/11 [1975] 1 CMLR D62 involving 'highly specialised trademarked goods', and *ARG/Unipart* [1988] OJ L 45/34 [1988] 4 CMLR 513.

[64] On market definition: Office of Fair Trading (2004d), National Economic Research Associates (1992), Areeda (1983), Baker and Bresnahan (1992), Kauper (1997), and Motta (2004: 102–115).

[65] Case 56/65 *Société Technique Minière v Maschinenbau Ulm GmbH* [1966] ECR 235, AG Opinion 257, Case T-4/89 *BASF AG v Commission* [1991] ECR II-1523, para 133, 231, Case T-8/89 *DSM NV v Commission* [1991] ECR II-1833, para 124, 221, *BMW Belgium* [1978] OJ L 46/33 [1978] 2 CMLR 126, para 139, Hildebrand (2002b: 188), and Gerber (2001: 291–292).

[66] Lutz (1989: 163). Also Beschle (1987), Landes and Posner (1981: 953), and Posner (2001: 17).

[67] Article 23 of Council Regulation (EC) No 1/2003 [2003] OJ L 1/1, Joined Cases 100–103/80 *Musique Diffusion Française SA v Commission* [1983] ECR 1825, para 106–108, Wils (2002a), and Smith (2001: para 9.06–9.07, 9.62).

[68] Gettier (1967) shows that, if attainable, knowledge of future contingent propositions must be more than a lucky or educated guess.

[69] Everitt and Fisher (1995: 52–56).

[70] Everitt and Fisher (1995: 20, 55–57).

[71] See references cited at note 65 above and Cases 56 and 58–64 *Établissements Consten Sàrl and Grundig-Verkaufs-GmbH v Commission* [1966] ECR 299, 340. Whether the potential effect is likely to be appreciable seems to turn on the ability of the firm to sustain the effect, Joined Cases

allocative inefficiency can be measured as discussed in Section III.A above, intervention on the basis of potential allocative inefficiency requires predictive techniques capable of justifying a belief that allocative inefficiency will occur due to the collusion. Predictive tools are appropriate when it is not possible or practical to observe the direct consequences of conduct.[72] Under the original system of *ex ante* enforcement, undertakings provided the Commission with information on which a prediction of allocative inefficiency could be based.[73] It has been doubted whether sufficient appropriate information was requested to enable an assessment of the applicability of Article 81(1) EC.[74] However, it is increasingly the role of economics in competition law to provide predictive tools capable of justifying a belief that allocative inefficiency is likely to be caused by the collusion under scrutiny.[75] Allocative inefficiency can be caused by collusion to contrive scarcity of output. This reduced output commands a higher price that more than compensates for lost sales.[76] Firms *able* to profitably contrive scarcity are said to possess market power and '[t]he economic concept of market power lies at the centre of the economic assessment of competition'.[77] Economics assumes that those *able* to profitably contrive scarcity *will* contrive scarcity. *Ex ante* intervention is justified when collusion will create market power or allow market power to be exercized. The problem is 'there is no *single* magnitude that comprehensively reflects market power'.[78] Community law has not adopted a particular market power test, and has instead relied on a number of economic approaches to predicting allocative inefficiency, depending on available data.[79] Four

T-374/94, T-375/94, T-384/94, T-388/94 *European Night Services v Commission* [1998] ECR II-3141, para 102, Case 30/78 *Distillers Company Limited v Commission* [1980] ECR 2229, para 28, and Case T-77/92 *Parker Pen Ltd v Commission* [1994] ECR II-549, para 46. Compare with Waelbroeck and Frignani (1999: 160–161).

[72] Patterson (2000: 435).

[73] Article 4(1) of Regulation 17. First Regulation Implementing Articles 81 and 82 of the Treaty [1962] OJ Special Edition 204/62, Green Paper on Vertical Restraints in EC Competition Policy Com (96) 721 Final [1997] 4 CMLR 519, para 188–192, Recitals of Regulation 2526/85 OJ [1985] L 240/1, Case 106/79 *Vereeniging ter Bevordering van de Belangen des Boekhandels v Eldi Records BV* [1980] ECR 1137, para 10, and Hildebrand (1998: 417–418).

[74] Kerse (1998: 72 n 18).

[75] Asch (1970, 1984 reprint: 131), Baker and Wu (1998: 1, 6–7), Bork (1993: 119–120), Jenny (1999: 192–198), Landes (number 45: 8), Markovits (1975a: 852), National Economic Research Associates (1992: 10–13), Nicolaides (2000: 12–16), Schmalensee (1982c: 25), and Sullivan (1977a: 1224).

[76] Van den Bergh and Camesasca (2001: 93).

[77] Bishop and Walker (1999: 27). On market power: Baker and Bresnahan (1992), Hay (1991), Klein (1993), Landes and Posner (1981), Office of Fair Trading (2004b), Price (1989) and Schmalensee (1982a)

[78] Asch (1970, 1984 reprint: 138, emphasis in original).

[79] For a list of considered factors see Guidelines on Vertical Restraints [2000] OJ C291/1, para 120–136.

factors examined to predict the allocative efficiency consequences of collusion are considered below. These are elasticity, the Lerner Index, market share and concentration, and barriers to market entry.

1. Own price elasticity

To cause allocative inefficiency, collusion must first enable *market* output to be reduced.[80] Second, the increased price must more than compensate for the lost sales. The second factor is dependent on the rate at which consumers cease buying the good when the price increases, and is measured by price elasticity (shown by the slope of the demand curve firms face). The more inelastic the demand curve the fewer the number of consumers that will cease buying the good for a given price hike and the greater the ability to profitably contrive scarcity.[81] Knowing the own price elasticity faced by a firm makes it relatively easy to determine *ex ante* whether output will fall and prices rise. However, the 'observation of actual elasticity magnitudes is ordinarily a complex and difficult task'.[82] Consequently, other economic tools are favoured in determining the ability to contrive scarcity.

2. Lerner Index

The Lerner Index identifies the extent to which market power has been exercized by looking at the difference between the cost of production (including normal profit) and price.[83] Firms able to price above cost in the long run have market power, since firms without market power can only make long-run normal profit. However, the variables required (price and marginal cost) are difficult to determine; even if this were not the case, the difficulty of separating those profiting from superior ability from those profiting from anti-competitive practices must be met.[84] Elzinga neatly sums up the validity of the index:

Joe S. Bain had little enthusiasm for the Lerner Index as a device for unmasking [market power]. He argued that gathering the measure's requisite data would be

[80] This aspect is considered in Section III.B.4. [81] Motta (2004: 106–107).

[82] Asch (1970, 1984 reprint: 141). Also Landes and Posner (1981: 957) and Posner (1976: 50). The use and difficulties with using elasticity are considered in Werden (1998: 364–384), Torre (2005), Levy (2003: para 8.04(3)–8.04(4), Van den Bergh and Canesasca (2001:93–94), Motta (2004: 107), Papandreou (1949), and Singer (1968: 70–72).

[83] Lerner (1934: 169–175), Landes and Posner (1981: 939–943, 960–961), Van den Bergh and Camesasca (2001: 94–95), Motta (2004: 116–117), Hylton (2003: 237–237), Bishop and Walker (1999: para 2.36), Posner (1976: 48, 56–57), and Neven *et al.* (1998: 53).

[84] Elzinga (1989: 26–27), Demsetz (1973: 20), and Bork (1993: 95–98, 125–127, 181).

'next to impossible.' In their assessment, however, Landes and Posner claim that the Lerner Index is 'a precise economic definition of market power.' Both characterizations of the index are on point. Precise it may be as a definition, practical it is not as a yardstick... Rarely has the Lerner Index been used. And this is probably just as well.[85]

3. Market share and concentration

Market share is probably the most used (though not necessarily the most useful) factor in considering the ability to contrive scarcity of output. As described by Landes and Posner:

[t]he standard method of proving market power... involves first defining a relevant market in which to compute the defendant's market share, next computing that share, and then deciding whether it is large enough to support an inference of the required degree of market power.[86]

In addition to market share, market concentration is also considered. As the number of firms competing in a market decreases, the demand faced by those remaining becomes more inelastic, giving the reduced number of rivals an increased ability to profitably contrive scarcity.[87] The remaining firms may have the power to reduce output alone or the smaller number of market participants may make collusion more feasible and likely.[88] The Court's concern with market share and concentration is evident in a number of important cases.[89] However, whilst market share and concentration *may* be useful for predicting anti-competitive outcomes, the predictions should be used with caution.[90] There is no clear link between market concentration

[85] Elzinga (1989: 27–28). Also Singer 1968: 64–65, Blair and Kaserman (1985: 110–112), Markovits (1975a: 844–845), and McGee (1971: 85–89).

[86] Landes and Posner (1981: 938, also 958–960) and American Bar Association (Antitrust Section) and Hartley (chairman) (1999: 108–113).

[87] Asch (1970, 1984 reprint: 148–158), Bishop and Walker (1999: para 2.44–2.46), Blair and Kaserman (1985: 234–237), Carlton and Perloff (2000: 251–259), Fox and Sullivan (1987: 942–944), Posner (1976: 51–55), and Kaysen (1951).

[88] Van den Bergh and Camesasca (2001: 89–92).

[89] Case C-306/96 *Javico International and Javico AG v Yves Saint Laurent Parfums SA* [1998] ECR I 1983, para 23–24, Case C-250/92 *Gottrup-Klim Grovvareforening and Others v Dansk Landbrugs Grovvareselskab Amba (DLG)* [1994] ECR I-5641, AG Opinion para 23, Case T-77/92 *Parker Pen Ltd v Commission* [1994] ECR II-549, para 46, and Faull and Nikpay (1999: para 1.10–1.11). Compare with Guidelines on the Assessment of Horizontal Mergers under the Council Regulation on the Control of Concentrations between Undertakings [2004] OJ C 31/5, para 14–21 and Varona *et al.* (2002: para 6.21–6.51).

[90] Landes and Posner (1981: 944–951), Schmalensee (1982a: 1798–1807), Lerner (1934: 165–168), and McGee (1971: 53–79).

and market output.[91] And '[s]ince a firm with a high market share can not safely be defined as one with market power, economists have restrained their enthusiasm for the measure'.[92]

4. Barriers to entry

A firm's ability to profitably contrive scarcity of output is affected by the speed and ability of new firms to supply the good. If new firms can enter the market and supply the good once scarcity is contrived, incumbent undertakings have little incentive to contrive scarcity, since they will simply lose sales.[93] Recognition of the relationship between barriers to entry and the ability to contrive scarcity of output is seen in the concern to ensure that parties can react to changing market conditions in the long run and the market is not foreclosed.[94] It is only when market access is impeded that there is a competition law problem.[95] As long as it remains possible for others to enter the market, it is unlikely that incumbents are able to contrive scarcity of output.[96]

What should be considered as improperly preventing entry is subject to debate.[97] The debate generally begins with a distinction between incumbent and entrant firms; barriers are seen as advantages incumbents have over potential entrants, reflected by an incumbent's ability to persistently charge more than the cost of production without attracting new competitors into the market.[98] Entry barriers may be structural, such as economies

[91] Demsetz (1973), Elzinga (1989: 14–16), Kahn (1953: 33–37), Muris (1998: 780), Muris (1997: 303–306), Stigler (1982: 7), and Van den Bergh and Camesasca (2001: 98–101, 170–191).

[92] Elzinga (1989: 16). Also Posner (1976: 55–56).

[93] Posner (1976: 49–50, 57–59).

[94] Case C-250/92 *Gottrup-Klim Grovvareforening and Others v Dansk Landbrugs Grovvareselskab Amba (DLG)* [1994] ECR I-5641, AG Opinion para 22, Case C-234/89 *Stergios Delimitis v Henninger Bräu AG* [1991] ECR I-935, para 19, 27, Case 23/67 *Brasserie de Haecht v Consorts Wilkin-Janssen* [1967] ECR 407, 415, Comanor and Frech III-1985), Blair and Kaserman (1985: 321–322), Faull and Nikpay (1999: para 7.102) and *Maison Jallatte* [1966] JO 37/66 [1966] CMLR D1, D3.

[95] Guidelines on Vertical Restraints [2000] OJ C 291/1, para 138, Blair and Kaserman (1985: 415), Dobson and Waterson (1996: 19), and Pheasant and Weston (1997).

[96] On entry barriers: Guidelines on the Assessment of Horizontal Mergers under the Council Regulation on the Control of Concentrations between Undertakings [2004] OJ C 31/5, para 68–75, and Varona *et al.* (2002: para 8.01–8.24).

[97] The hallmark of Chicago School analysis is the belief that there are few barriers to entry: Posner (1979a: 944–946).

[98] Bain (1993: 3). Also Carlton and Perloff (2000: 77) and Stigler (1983c: 67). A problem with this definition is that it fails to find barriers when supra-normal profits are dissipated by rent-seeking, hence Demsetz (1982: 48–49) argues that a definition of barriers focusing on asymmetry between entrants and incumbents cannot identify all barriers.

of scale and cost advantages.[99] The extent to which structural barriers are relevant to an Article 81(1) EC competition enquiry is contestable, since it can be argued that nothing should be characterized as a barrier if the market situation would be no better, or worse, in the absence of the barrier.[100] Structural barriers are distinct from strategic barriers, which involve efforts by incumbents to keep potential competitors out of the marketplace.[101] As Comanor observes, to undertakings 'entry barriers are an asset which has value similar to that of a new machine or a well-received trademark, and, therefore, firms can be expected to adopt policies which can be explained only as an "investment in entry barriers"'.[102] Various strategies may be used to prevent new firm entry, eg, raising the costs potential entrants must meet to enter the market or altering the market structure in a manner unfavourable to potential entrants.[103] Investment in capacity, advertising, choice of location, research and development, and vertical integration can all be used to deter entry.[104] The difficulty is distinguishing strategic investment from investment that would have occurred in the absence of the barricading strategy.[105] In some cases, there may be evidence indicating that investment is strategic.[106] In the absence of an admission or evidence of strategic purpose, some simply presume strategic use if the agreement is of an especially long duration.[107] Peeperkorn reflects this view by writing that whilst foreclosure 'may be mitigated by stronger ex-ante competition between suppliers to obtain single-branding contracts,... the longer the duration the more likely it will be that this effect will not be enough to fully compensate for the lack of inter-brand competition'.[108]

[99] Bain (1993: 54–55, 98–109, 114–116, 144–147), Viscusi *et al.* (2000: 158–159), Carlton and Perloff (2000: 79), Geroski and Schwalbach (1989: 9–10, 13–15), and Odudu (2002b: 471–474).

[100] Bork (1993: 195), Chamberlin (1950), and Demsetz (1982: 50–53).

[101] London Economics (1994: 7). It is noted that there is a degree of interdependence between structural and strategic barriers, since structural barriers may be relevant to assessing the credibility of strategic barriers: Encaoua *et al.* (1986: 61–69).

[102] Comanor (1967: 261).

[103] Salop and Scheffman (1983), Encaoua *et al.* (1986), Viscusi *et al.* 2000: 170–186), Dixit (1980), Lieberman (1987), Spence (1977), Milgrom and Roberts (1982), Fudenberg and Tirole (1986), and Economist (2004).

[104] Encaoua *et al.* (1986: 56–64), Salop (1979), Salop (1981), Van den Bergh and Camesasca (2001: 119), Mathewson and Winter (1987), Patterson (2000: 437–445), Patterson (1997), Schwartz (1987), Williamson (1979b: 960), and Williamson (1987: 274).

[105] The distinction turns on purpose/intent: Salop (1979: 335) and Encaoua *et al.* (1986: 57).

[106] Consider *Van Den Bergh Foods* [1998] OJ L 246/1 [1998] 5 CMLR 530, para 64–68.

[107] Blair and Kaserman (1985: 415–417) and Van den Bergh and Camesasca (2001: 224). Qualified by Aghion and Bolton (1987).

[108] Peeperkorn (1998: 12). Consider also Article 5(a) of Commission Regulation (EC) No 2790/1999 [1999] OJ L 336/21, Guidelines on Vertical Restraints [2000] OJ C 291/1, para 57–60, Whish (2000b: 917–918), and Korah and O'Sullivan (2002: para 5.1.1).

C. *Presumptions*

Allocative inefficiency can be identified by both *ex post* measurement and *ex ante* prediction, and both approaches are used to show the *effect* of restricting competition within Article 81(1) EC.[109] However, measurement and prediction may not identify, or may not be the most expeditious way to identify, conduct causing allocative inefficiency.[110] Thus, Article 81(1) EC can be applied absent measured inefficiency or inefficiency predicted using economic tools. Such intervention is permissible when collusion has the *object* of restricting competition. This is clearly stated by the Court in *Consten and Grundig*, finding that 'for the purpose of applying Article [81(1) EC], there is no need to take account of the concrete *effects* of an agreement once it appears that it has as its *object* the prevention, restriction or distortion of competition'.[111] This is most clearly confirmed in *Ferriere Nord*, when Ferriere Nord sought to rely on the Italian version of the EC Treaty, which 'unlike the other language versions, refers to agreements which have as their object *and* effect the restriction of competition', to argue that the 'provision lays down a cumulative and not an alternative condition'.[112] Rejecting the claim, the Court forcefully confirms that the object and effect conditions provide distinct tests.[113]

Whilst the object and effect tests are distinct, they both aim at identifying the same consequence of collusion—allocative inefficiency. Thus in *Société Technique Minière* the Court confirms that 'Article [81(1)] is based on an assessment of the effects of an agreement from two angles of economic evaluation'.[114] The task is to justify competition law intervention when

[109] Compare Guerrin and Kyriazis (1993: 800–802).

[110] Harding and Joshua (2003: 144, 151, 158–159), Asch (1970, 1984 reprint: 5), Hildebrand (1998: 135–136, 242), and Gerber (2001: 56).

[111] Cases 56 and 58–64 *Établissements Consten Sàrl and Grundig-Verkaufs-GmbH v Commission* [1966] ECR 299, 342, emphasis added. Also Black (2005: 115–126) and Robertson (2002: 96–97).

[112] Case C-219/95 P *Ferriere Nord SpA v Commission* [1997] ECR I-4411, AG Opinion para 12, emphasis in original.

[113] Case C-219/95 P *Ferriere Nord SpA v Commission* [1997] ECR I-4411, para 14–16, AG Opinion para 15–18, Case T-143/89 *Ferriere Nord SpA v Commission* [1995] ECR II-917, para 31, Case 56/65 *Société Technique Minière v Maschinenbau Ulm GmbH* [1966] ECR 235, 249, Cases 56 and 58–64 *Établissements Consten Sàrl and Grundig-Verkaufs-GmbH v Commission* [1966] ECR 299, 343, Case 123/83 *Bureau National Interprofessional du Cognac v Guy Clair* [1985] ECR 391, AG Opinion 398–399, Case C-250/92 *Gottrup-Klim Grovvareforening and Others v Dansk Landbrugs Grovvareselskab Amba (DLG)* [1994] ECR I-5641, AG Opinion para 16, Joined cases T-374/94, T-375/94, T-384/94, T-388/94 *European Night Services v Commission* [1998] ECR II-3141, para 136, and Lasok (1982: 136).

[114] Case 56/65 *Société Technique Minière v Maschinenbau Ulm GmbH* [1966] ECR 235, 248. This rejects the 'distinctness' thesis as articulated in Black (2005: 95, 96, 115, 116–119, 121, 126), which argues that the tests identify different things.

inefficiency is neither measured nor predicted by economic tools.[115] It can be shown that in addition to measured and predicted allocative inefficiency, intervention is based on legal presumptions of allocative inefficiency.[116] These presumptions are established in two ways.

First, presumptions are established inductively once there is sufficient experience of particular conduct and of its impact on allocative efficiency. Second, there is a presumption that if an outcome is intended it is more likely to occur than if that same outcome is not intended. Legal presumptions, based on experience and intent, fit with the Court's instruction that determining the object of an agreement is 'a question of examining the aims pursued by the agreement as such, in the light of the economic context in which the agreement is to be applied.'[117] The basis and content of legal presumptions of allocative inefficiency are set out below.

1. Induction based presumptions

Goyder writes that 'some contractual restrictions are, prima facie, so likely to affect competition that this effect will be presumed'.[118] The welfare consequences of some types of collusion are so well known that there is a legal presumption of their allocative inefficiency. This follows US antitrust law, where there is a legal presumption that an anti-competitive outcome will occur from what is referred to as *per se* conduct. In *Northern Pacific Railway Co v United States*, Justice Black delivered the opinion of the Supreme Court, stating that:

> there are certain agreements or practices which *because of their pernicious effect on competition and lack of any redeeming virtue are conclusively presumed to be unreasonable* and therefore illegal without elaborate inquiry as to the precise harm they have caused or the business excuse for their use.[119]

[115] On avoiding an 'effect' type enquiry in US antitrust: Calkins (2000).

[116] Guidelines on the Application of Article 81(3) of the Treaty [2004] OJ C 101/97, para 21, 24.

[117] Joined Cases 29/83, 30/83 *Compagnie Royale Asturienne des Mines SA and Rheinzink GmbH v Commission* [1984] ECR 1679, para 26.

[118] Goyder (1998: 121). Faull and Nikpay (1999: para 2.61) speak of the practice's 'natural tendency'. Rose (1993: para 2-071, 2-097, para 2-100) speaks of 'the obvious consequence of the agreement.' Compare with Guidelines on the Application of Article 81(3) of the Treaty [2004] OJ C 101/97, para 21.

[119] *Northern Pacific Railway Co v United States* 356 US 1, 5 (1957), emphasis added. The irrebuttable nature of the presumption is in decline: *California Dental Association v Federal Trade Commission* 526 US 756, 780–781 (1999), American Bar Association (Antitrust Section) and Hartley (chairman) (1999: 12–15, 128–131, 135–139, 162–163), Areeda (1986: 583), Beschle (1987: 482–484, 486–496), Collins (1999), Keyte (1997), Liebeler (1983: 392–396), and Meese (2000: 478–489).

The presumption is based on the foreseeable consequences of the collusion; foreseeability is established inductively so that:

[p]roperly understood, *per se* rules develop from a rule-of-reason analysis ... With the passage of time the courts came to identify certain types of agreement which are conclusively presumed to be 'without redeeming virtue'... obviating the need for more detailed market investigation.[120]

The Commission justifies presumption based intervention on the basis that:

a full economic analysis of every case would be very costly and might not be justified by gains in identifying market situations ... that were detrimental to competition. In those circumstances, competition policy may have to resort to relatively simple rules of thumb and do without a full economic analysis of every case.[121]

Measurement and prediction are avoided because such an analysis would only confirm what logic and experience tell us is a virtually certain outcome.[122] A legal presumption spares the expense of an 'effect' enquiry and has clear deterrent value because of the absolute nature of the prohibition.[123] The validity of these claims depends on the content of the category and the nature of the presumption.[124]

(a) The content of the category The text of Article 81(1) EC gives guidance on the types of collusion presumed to result in allocative inefficiency.[125] The CFI has identified a narrower list of conclusive legal presumptions of allocative inefficiency, consisting of '*obvious restrictions of*

[120] Craig and de Búrca (2003: 952). Also Black (1997: 149 note 27). Compare with the received view in Black (2005: 96, 119–127). Burnet (1984: 3–7) rejects the inductive development thesis. Goyder (1998: 565–566) describes the inductive development of the approach to Article 81(3) EC.

[121] Green Paper on Vertical Restraints in EC Competition Policy Com (96) 721 Final [1997] 4 CMLR 519, para 86. Also Communication from the Commission on the Application of the Community Competition Rules to Vertical Restraints (Follow-up to the Green Paper on Vertical Restraints) Com(98)544 Final [1998] OJ C 365/3, 9, 21, Easterbrook (1984: 9–10), and Neven *et al.* (1998: 17–19). Korah and O'Sullivan (2002: para 8.6) 'would prefer a full economic appraisal of each agreement to be made'.

[122] Webb (1983: 708–710), American Bar Association (Antitrust Section) and Hartley (chairman) (1999: 2–3), and Black (1997: 151–152).

[123] American Bar Association (Antitrust Section) and Hartley (chairman) (1999: 2–3), Hylton (2003: 116), Webb (1983: 708–710), Black (1997: 151–152).

[124] On the trade-off between legal certainty created by presumptions and economic reality see Terhorst (2000: 373–377) and Korah and O'Sullivan (2002: para 3.4.13.7, 8.2).

[125] These are: (a) directly or indirectly fix purchase or selling prices or any other trading conditions; (b) limit or control production, markets, technical development, or investment; (c) share markets or sources of supply; (d) apply dissimilar conditions to equivalent transactions with

competition such as price-fixing, market-sharing or the control of outlets'.[126] This describes what is increasingly termed hard-core cartel conduct, about which '[t]here is now little disagreement in terms of competition theory and policy at both international and national levels about the damaging effect of such arrangements on public and consumer interests'.[127] Faull and Nikpay write that:

> the undisputed illegality in the Community of the so-called hard core cartels is shown by the fact that, in the vast majority of appeals against Commission decisions that condemned them, the issues at stake were the standard of proof and the amount of the fines imposed on the undertakings.[128]

Much more controversial is whether there is any role for legal presumptions of allocative inefficiency in relation to vertical collusion.[129] Undertakings may enter vertical arrangements to contrive scarcity and increase price.[130] One view is that a vertical fixed minimum price facilitates horizontal collusion between suppliers.[131] By rendering pricing transparent, cheating on the agreed price is easier to detect. More importantly, because the buyer cannot pass any discount they received to the consumer and win more business, the supplier has no incentive to reduce the price charged to buyer.[132] Van den Bergh and Camesasca feel that less than a third of vertical minimum price fixing cases involve the hint of a cartel and doubt whether the price restraint in itself is sufficient to support a cartel.[133] A second view of vertical minimum prices is that they disguise horizontal price fixing

other trading parties, thereby placing them at a competitive disadvantage; (e) make the conclusion of contracts subject to acceptance by the other parties of supplementary obligations which, by their nature or according to commercial usage, have no connection with the subject of such contracts.

[126] Joined cases T-374/94, T-375/94, T-384/94, T-388/94 *European Night Services v Commission* [1998] ECR II-3141, para 136, emphasis added.

[127] Harding and Joshua (2003: 1). Also Guidelines on the Applicability of Article 81 of the EC Treaty to Horizontal Cooperation Agreements [2001] OJ C 3/2, para 19, 25, Guerrin and Kyriazis (1993: 797–800), Jephcott and Lübbig (2003: vii), Blair and Kaserman (1985: 164–169), Faull and Nikpay (1999: para 6.06–6.33), Whish (2003: 454), and Gerber (1994: 111–113, 121–122).

[128] Faull and Nikpay (1999: para 6.08).

[129] Vertical relationships are between producers selling complementary rather than competing (substitute) goods and services: Article 2(1) of Commission Regulation (EC) No 2790/1999 [1999] OJ L 336/21, Guidelines on Vertical Restraints [2000] OJ C 291/1, para 23–24, 100, and Carlton and Perloff (2000: 396).

[130] Carlton and Perloff (2000: 379, 387–395).

[131] Carlton and Perloff (2000: 409), Gerhart (1981: 423), Baker (1996: 520–523), and Liebeler (1983: 404–405).

[132] Telser (1960: 96–104), Van den Bergh and Camesasca (2001: 216), and Blair and Kaserman (1985: 354–355).

[133] Van den Bergh and Camesasca (2001: 223) and Butz and Kleit (2001).

amongst buyers. The supplier is induced to impose the price restraint, making cheating easier to detect.[134] Even in the absence of a cartel, price-competition amongst buyers of the contract goods is eliminated.[135] The problem with the buyer collusion theory is that collusion by buyers conflicts with the suppliers' interest of maximizing sales; there is no incentive for the supplier to police a buyers' horizontal price fixing agreement and a strong incentive to do just the opposite.[136] Further, unless there are barriers to entry at the buyer level, the super-normal profits being made will attract new entrants who will erode the collusive price.[137]

Despite the risks to competition, it is generally recognized that vertical collusion can be viewed as a cost-effective way for undertakings to get goods and services to the end consumer.[138] For example, vertical maximum price fixing can be used to increase output and reduce prices when there is a problem of double marginalization. If both supplier-undertakings and buyer-undertakings enjoy a degree of market power, the supplier exercizes market power against the buyer, charging more and selling a reduced quantity. The buyer then exercizes market power in relation to the end consumer, further reducing output and increasing the price paid: the end consumer pays two monopoly additions.[139] Both consumers and the supplier benefit from a maximum resale price because the buyer-undertaking is limited in the amount of market power that can be exercised, increasing the quantity of goods offered to the end consumer and lowering the price.[140] The double-edged nature and ambiguity in operation makes vertical collusion difficult to regulate and forces Goyder to reject a view that enforcing competition law simply requires 'a firm hand and a cool nerve

[134] Guidelines on Vertical Restraints [2000] OJ C 291/1, para 112, Telser (1960: 104), Blair and Kaserman (1985: 353–354), Tirole (1988: 184–185), Dobson and Waterson (1996: 24–25), Gerhart (1981: 423–424), and Liebeler (1983: 405–406).

[135] Guidelines on Vertical Restraints [2000] OJ C 291/1, para 112.

[136] Kelly (1988: 368–370).

[137] Van den Bergh and Camesasca (2001: 216–217).

[138] Coase (1937), Adelman (1949), Blair and Kaserman (1985: 285–295), Boscheck (2000: 18–20), Comanor and Frech III (1985: 545), Mathewson and Winter (1987), Dobson and Waterson (1996), Neven *et al.* (1998: 18–24), Green Paper on Vertical Restraints in EC Competition Policy Com (96) 721 Final [1997] 4 CMLR 519, executive summary para 10 and Carlton and Perloff (2000: 378, 396).

[139] Tirole (1988: 174–176), Van den Bergh and Camesasca (2001: 217), Carlton and Perloff (2000: 398–400), and Dobson and Waterson (1996: 7–8).

[140] Tirole (1988: 177), Bork (1993: 280–298), Van den Bergh and Camesasca (2001: 217–218), Grimes (1998: 604–606), Massey (1998: 5), Blair and Kaserman (1985: 297–299), Carlton and Perloff (2000: 390–392, 400). For critique of the double marginalization justification: Blair and Kaserman (1985: 295–304), Carlton and Perloff (2000: 400), Comanor (1967: 257–259), and Grimes (1998: 580–583), Schmalensee (1973: 444–448), Van den Bergh and Camesasca (2001: 226), Vernon and Graham (1971), Warren-Boulton (1974), and Westfield (1981).

with little sympathy for those who have found themselves caught up in its operations'.[141]

The potential for efficiency gains makes it difficult to identify vertical conduct that ought to be presumed to cause allocative inefficiency.[142] However, the efficiencies possessed by some type of vertical collusion seem to be productive rather than allocative. In defending vertical collusion, Neven, Papandropoulos and Seabright write that 'there are many circumstances under which they enhance *productive efficiency*'.[143] As Van den Bergh and Camesasca write:

> courts and policy makers sometimes phrase the evaluation of a vertical restraint in terms of the promotion of inter-brand competition versus the restriction of intra-brand competition. However, this is not the relevant balancing act. Benefits in terms of efficiency savings must be weighed against costs in terms of restrictions of competition.[144]

The meaning of restricting competition advocated—allocative inefficiency—means that productive efficiencies are not relevant to the Article 81(1) EC enquiry.[145] Thus, Article 4 of Regulation 2790/1999 reflects a legal presumption of allocative inefficiency in relation to vertical minimum price-fixing, absolute territorial protection, and certain customer restraints.[146] Faull and Nikpay describe these practices as having the object

[141] Goyder (1993: 2).

[142] Van den Bergh and Camesasca (2001: 215, 247), Case 56/65 *Société Technique Minière v Maschinenbau Ulm GmbH* [1966] ECR 235, 248–249, Case 32/65 *Italian Republic v Council of the EEC and Commission of the EEC* [1966] ECR 389, AG Opinion 419, Schaub (1999: 143), Green Paper on Vertical Restraints in EC Competition Policy Com (96) 721 Final [1997] 4 CMLR 519, American Bar Association (Antitrust Section) and Hartley (chairman) (1999: 12, 105–107), Boscheck (2000: 18–20), Comanor and Frech III (1985: 545), Hawk (1988: 66, 74–75), Mathewson and Winter (1987), Neven *et al.* (1998: 18–24), and Nicolaides (2000: 10).

[143] Neven *et al.* (1998: 18), emphasis added. Also Guidelines on Vertical Restraints [2000] OJ C291/1, para 115–118, Comanor (1985: 989, 991), Gould and Yamey (1967: 729), Bork (1967b: 742–743), Van den Bergh and Camesasca (2001: 252), Gerhart (1981: 439), Comanor (1985: 989–991), Gould and Yamey (1967: 729), Bork (1967b: 742–743), Liebeler (1982: 5), Griffiths (2000: 245), Whish (2001: 542), Carlin (1996: 284), Korah and O'Sullivan (2002: 26), Carlton and Perloff (2000: 401–405), Blair and Kaserman (1985: 351–353), Dobson and Waterson (1996: 8–9), Korah and O'Sullivan (2002: 28–37), Gerhart (1981: 439), Telser (1960: 89–96), Tirole (1988: 182–183), Van den Bergh and Camesasca (2001: 219), Klein and Murphy (1988), and Hylton (2003: 113–116).

[144] Van den Bergh and Camesasca (2001: 252).

[145] In Chapter 6 it will be argued that the concern of Article 81(3) EC is with productive efficiency.

[146] Commission Regulation (EC) No 2790/1999 [1999] OJ L 336/21, Guidelines on Vertical Restraints [2000] OJ C291/1, para 7, 46–56, Whish (2000b: 911–917). It is important to note that the block exemption reflects rather than establishes a presumption of allocative inefficiency.

of restricting competition.[147] The positive allocative efficiency aspect of vertical maximum price fixing is recognized by Article 4(a) of Regulation 2790/1999 and the Commission Guidelines on Vertical Restraints, which expressly provide that vertical maximum retail pricing is not treated as a hardcore restraint.[148] The pricing strategy is not *presumed* to cause allocative inefficiency.

Categories of conduct presumed to cause allocative inefficiency exist.[149] However, conduct must still be classified in subsequent cases. Much energy can be devoted to classification, and arguments advanced as to why collusion does or does not fit within a particular category.[150] The effort devoted to this task may well erode all savings that justify the existence of legal presumptions.[151] However, the incentive to make such effort can be minimized, or the effort redirected, if the presumptions are rebuttable rather than conclusive.

(b) Nature of the presumption There is a legal presumption that certain types of collusive conduct cause allocative inefficiency. A pressing question is whether the presumption is rebuttable or conclusive. *Consten and Grundig* suggests the presumption is conclusive. Once the presumption exists, '[n]o further considerations ... can in any way lead ... to a different solution under Article 81(1)'.[152] *Société Technique Minière* suggests the presumption is rebuttable, it being impossible to make 'any kind of advance

There is a view that the existence of a block exemption regulation implies that all conduct falling within the scope of the regulation also falls within the scope of Article 81(1) EC: Case 32/65 *Italian Republic v Council of the EEC and Commission of the EEC* [1966] ECR 389, 405, Bovis (2001: 98), Forrester QC (2001: 173), Forrester and Norall (1984: 15, 31, Jenny (1999: 194), Jones and Sufrin (2001: 519), and Lever QC and Peretz (1999: para 2.2–2.7). This view is rejected: Case 32/65 *Italian Republic v Council of the EEC and Commission of the EEC* [1966] ECR 389, 405–406, AG Opinion 416 and Whish (2001: 546). A block exemption regulation does not and cannot determine whether Article 81(1) EC is infringed.

[147] Faull and Nikpay (1999: para 7.25).

[148] Guidelines on Vertical Restraints [2000] OJ C 291/1, para 225–228, Whish (2000b: 912), and Subiotto and Amato (2001: 168).

[149] Whish (2003: 112) and Korah and O'Sullivan (2002: 104–108).

[150] Guidelines on Vertical Restraints [2000] OJ C 291/1, para 47, Keyte (1997), American Bar Association (Antitrust Section) and Hartley (chairman) (1999: 128–131, 135–139, 162–163), Beschle (1987: 482–496), and Areeda (1986).

[151] Beschle (1987: 473, 486–496) and American Bar Association (Antitrust Section) and Hartley (chairman) (1999: 4, 9).

[152] Cases 56 and 58–64 *Consten and Grundig* [1966] ECR 299, 343. Also Case 123/83 *Bureau National Interprofessional du Cognac v Guy Clair* [1985] ECR 391, AG Opinion 398–399, and Case C-250/92 *Gottrup-Klim Grovvareforening and Others v Dansk Landbrugs Grovvareselskab AmbA (DLG)* [1994] ECR I-5641, AG Opinion para 16.

judgment with regard to a category of agreements determined by their legal nature'.[153]

Irrebuttable presumptions of allocative inefficiency remain unproblematic if established in a manner that keeps false positives to a minimum.[154] However, measurement and prediction are avoided only because such an analysis would only confirm what has been inductively determined. Inductive knowledge is accorded no normative value unless capable of falsification and subject to testing; it is the durability of a falsifiable claim that makes it valuable.[155] The presumption must be rebuttable.[156] If the presumption is to be rebutted, this will require showing that allocative inefficiency is not caused by conduct presumed to cause allocative inefficiency. Measurement and prediction are required—the *effect* analysis cannot be avoided.[157] The advantage of legal presumptions—a cheaper, quicker and less complicated way to determine allocative efficiency consequences than measurement or prediction—is lost.[158] The legal presumption is not without value. It operates as a burden-shifting tool, forcing undertakings to demonstrate the absence of detrimental consequences.[159]

2. *Intent based presumptions*

Goyder writes that within Article 81(1) EC '[t]he agreement must either be *intended* to have such a result on competition or regardless of the parties'

[153] Case 56/65 *Société Technique Minière v Maschinenbau Ulm GmbH* [1966] ECR 235, 248. Also Case 32/65 *Italian Republic v Council of the EEC and Commission of the EEC* [1966] ECR 389, AG Opinion 419, Schaub (1999: 143), and American Bar Association (Antitrust Section) and Hartley (chairman) (1999: 1).

[154] Hylton (2003: 130–131) and American Bar Association (Antitrust Section) and Hartley (chairman) (1999: 3–4). By false positives it is meant collusion presumed to cause, but not actually causing, allocative inefficiency.

[155] Popper (2002: 3–26, 57–73, 312–318) and Caldwell (1991: 2–13).

[156] See references cited at note 119 above, Easterbrook (1984: 10, 19), American Bar Association (Antitrust Section) and Hartley (chairman) 1999: 142–143). Case 262/81 *Coditel SA v Ciné-Vog Films SA* [1982] ECR 3381, Case 27/87 *Louis Erauw-Jacquery Sprl v La Hesbignonne Sc* [1988] ECR 1919 and Case 26/76 *Metro SB-Großmärkte GmbH & Co KG v Commission* [1977] ECR 1875 may provide examples.

[157] Compare with American Bar Association (Antitrust Section) and Hartley (chairman) (1999: 12–15), Areeda (1986: 583), Liebeler (1983: 392–396), and Collins (1999).

[158] Piraino Jr (1994: 1761–1763) and Burnet (1984: 11–13).

[159] The allocation and shifting of the burden of proof in US antitrust is discussed in Beschle (1987: 504–506), Calkins (2000: 520–521), Carrier (1999), Piraino Jr (1994), and Piraino Jr (1991). On the burden of proof: Article 2 of Council Regulation (EC) No 1/2003 [2003] OJ L 1/1, Cases 6–7/73 *Istituto Chemioterapico Italiano SpA et Commercial Solvents Corporation v Commission* [1974] ECR 223, AG Opinion 269, Brealey (1985), Ortiz Blanco (1996: 44–45, 102), Smith (2001: para 10.30–10.32).

intentions, actually have such a result'.[160] That intention plays a role in the 'object' approach is acknowledged.[161] It can be argued that intent is relevant because undertakings colluding with the intention of restricting competition are more likely to restrict competition than if they had no such intention.[162] Norrie opines that when we say that a person intends we simply mean that they see it as the *virtually certain* result of their action, ie, we are making a prediction of outcome.[163] In the same vein, Kenny has written that the law is concerned with intention to achieve outcomes because outcomes are more likely when agents are working to achieve them.[164] Bork writes that '[i]mproper exclusion (exclusion not the result of superior efficiency) is always deliberately intended'.[165] Those intending to deny consumers the benefits of competition, left unabated, will continue until their scheme is successful, as seen in *Addyston Pipe & Steel v United States*: having tried but failed to raise prices, the companies reported that 'the system . . . has not, in its operation, resulted in the advancement in the prices of pipe, as was anticipated . . . and *some further action is imperatively necessary in order to accomplish the ends for which this association was formed*'.[166]

Intent's predictive nature justifies a legal presumption that collusion with the intention to contrive a scarcity of output will lead to contrived scarcity of output. The existence of such a legal presumption seems clear in US antitrust.[167] In *Chicago Board of Trade v United States*, Justice Brandeis felt it

[160] Goyder (1998: 120), emphasis added. Also Case T-17/93 *Matra Hachette SA v Commission* [1994] ECR 593, para 85. Compare the approach of the following commentators: Bishop and Walker (1999: 122–132), Cunningham (1973: 51), Deringer (1968: para 131), Fitzsimmons (1994: 21–22), Furse (2000: 119–121), Joliet (1967: 133–136), Ritter *et al.* (2000: 92–94), Rodger and MacCulloch (1999: 138), Toepke (1982: § 222.4), Wesseling (2000: 28–29) and *David John Passmore v Morland* [1999] European Commercial Cases 367 para 12.

[161] Case C-551/03 P *General Motors BV and Opel Nederland BV*, AG Opinion para 64, 77–79 and Guidelines on the Application of Article 81(3) of the Treaty [2004] OJ C 101/97, para 22. The editors of the Common Market Law Reports initially interpreted Article 81(1) EC to prohibit agreements '*designed* to prevent, restrict or distort competition within the Common Market or which have this *effect*'. This interpretation, in which object is replaced by designed, suggests that rather than being a term of art, object is synonymous with specific intent, purpose, and design. Intent and purpose are used interchangeably. See Case 13/61 *Robert Bosch GmbH and another v Kleding-Verkoopbedrijf de Geus en Uitdenbogerd* [1962] CMLR 1 at note 2 and Chantry (1991: 43–46).

[162] *Lessig v Tidewater Oil Co* 327 F 2d 459, 474 (1964), Lever and Neubauer (2000: 8, 20), Sullivan 1977a: (1229–1231), Sullivan (1977b: 166), Givens (1983: para 3.09), Cass and Hylton (2001: 679–707, 712–715) and Gifford (1986: 1045). The validity of this claim in relation to Article 81(1) EC is questioned in Black (2005: ch 4).

[163] Norrie (1989: 800–802).

[164] Kenny (1966: 649–651).

[165] Bork (1993: 160). Also Waller (2001: 334).

[166] *United States v Addyston Pipe & Steel* 85 F 271, 274 (1898), emphasis added.

[167] Odudu (2002b), and American Bar Association (Antitrust Section) and Hartley (chairman) (1999: 114–116, 166–167). Compare with the use of intent in Kahn (1953: 48–53) and Markovits (1975a: 843).

important to consider 'the purpose or end sought to be attained'.[168] An enquiry into intent was necessary, 'not because a good intention will save an otherwise objectionable regulation or the reverse; but because *knowledge of intent* may help the court to interpret facts and *to predict consequences*'.[169] In *Swift & Company*, addressing the argument that 'intent can make no difference', Justice Holmes felt that '[w]here acts are not sufficient in themselves to produce a result which the law seeks to prevent . . . an intent to bring it to pass is necessary in order to produce *a dangerous probability that it will happen*'.[170] In other cases Chief Justice Hughes said 'knowledge of actual intent is an aid in the interpretation of facts *and prediction of consequences*', whilst Justice White felt an enquiry into purpose was relevant 'because it tends to show effect'.[171]

A legal presumption that collusion intended to restricted competition has led or will lead to a restriction of competition also seems to exist in EC competition law.[172] In *Consten and Grundig*, the Court prohibited an agreement after finding that '[t]he applicants thus *wished* to eliminate any possibility of competition'.[173] In *Hoffmann-La Roche*, the Court found that the conduct examined 'proves the *intention* to restrict competition'.[174] In *IAZ International Belgium* the Court condemned an agreement because '*the purpose of the agreement*, regard being had to its terms, the legal and economic context in which it was concluded and the conduct of the parties, *is appreciably to restrict competition within the common market*'.[175] In *Rhône-Poulenc* the applicant accepted that an 'infringement exists once the undertakings have committed themselves, *even if the commitment is simply mental* on the part of the undertakings' representatives and *even if it is not reflected on the market by anti-competitive conduct*'.[176]

[168] *Board of Trade of City of Chicago v United States* 246 US 231, 238 (1918), emphasis added.

[169] *Board of Trade of City of Chicago v United States* 246 US 231, 238 (1918), emphasis added.

[170] *Swift & Company v United States* 196 US 375, 396 (1905), emphasis added.

[171] *Appalachian Coals v United States* 288 US 344, 372 (1933) and *Broadcast Music Inc v Columbia Broadcasting System Inc and American Society of Composers, Authors and Publishers v Columbia Broadcasting System Inc* 441 US 1, 19–20 (1979). Also *Utah Pie Company v Continental Baking Company* 386 US 685, 702, 696–697 (1967).

[172] For earlier attempts to establish this proposition see Odudu (2001a) and Odudu (2001b).

[173] Cases 56 and 58–64 *Établissements Consten Sàrl and Grundig-Verkaufs-GmbH v Commission* [1966] ECR 299, 342–343, emphasis added. Verouden (2003: 543–544).

[174] Case 85/76 *Hoffmann-La Roche & Co AG v Commission* [1979] ECR 461, para 115, emphasis added.

[175] Joined Cases 96–102, 104, 105, 108 and 110/1982 *IAZ International Belgium NV v Commission* [1983] ECR 3369, para 25, emphasis added. Also Case Note (1993: 250).

[176] Case T-1/89 *Rhône-Poulenc SA v Commission* [1991] ECR II-867, para 107. Also Case T-8/89 *DSM NV v Commission* [1991] ECR II-1833, para 214.

A presumption exists that collusion intended to restrict competition results in, or will result in, a restriction of competition. Two challenges exist to making this presumption administrable. First, intent must be determined. Second, the nature of the presumption must be specified.

(a) Determining intent The first task is to determine the undertaking's intent. In *Distillers*, 'DCL admitted that their price terms *were intended ... to protect DCL sole distributors from competition*'.[177] However, if intention was subjectively determined 'it would be easy to plead [a contrary intention] dishonestly and difficult to rebut the plea. Indeed it might be well nigh impossible'.[178] According to Hildebrand, '[a]greements and decisions are often explicit enough for an anti-competitive purpose to be evident'.[179] This requires an objective method, so that intent is determined from external manifestations.[180] The objective method is essential, since we are considering the intention of undertakings, which have no subjective thoughts.[181]

(b) Nature of the presumption There is a legal presumption that collusion intended to restrict competition has restricted or will restrict competition based on the predictive qualities of intention.[182] As with presumptions based on experience, it must be asked whether the presumption is rebuttable or conclusive.[183] This depends on whether intervention is *ex post* or *ex ante*.

(i) Ex post If the collusive conduct is implemented, there is a question of whether measurement can be used to rebut the legal presumption of allocative inefficiency based on the predictive quality of intent. Measurement may be carried out which shows that collusion, though intended to cause allocative inefficiency, did not in fact cause allocative inefficiency.[184] Measurement would show an unsuccessful attempt. However, an unsuccessful attempt is not harmless. Posner recognizes that:

an opportunity to obtain a lucrative transfer payment in the form of monopoly profits will attract real resources into efforts by sellers to monopolize ... The cost

[177] *Distillers Company Limited* [1978] OJ L 50/16 [1978] 1 CMLR 400, para 85, emphasis added, also para 89.

[178] Denning (1961: 10–11).

[179] Hildebrand (2002b: 187).

[180] Denning (1961: 18), Odudu (2002b: 478–480), Odudu (2001a: 69–70), American Bar Association (Antitrust Section) and Hartley (chairman) (1999: 144–145), Gifford (1986: 1023–1024), Cass and Hylton (2001: 659, 664–665, 675–678), and Bork (1993: 36–37, 123, 157).

[181] On corporate intention: Slapper and Tombs (1999: 26–35), Laufer (1992: 1053–1058, 1085–1091), Laufer (1994), and Quaid (1998).

[182] See text accompanying notes 162–166.

[183] See Section III.C.1. (b).

[184] See Section III.A above.

of the resources so used are costs of monopoly just as much as the costs resulting from the substitution of products that costs society more to produce than the monopolized product.[185]

Attempting to contrive a scarcity of output has costs; these costs are said to place 'the economist's hostility to monopoly on somewhat firmer ground'.[186] The resources consumed in rent-seeking behaviour are perhaps the major cost of anti-competitive conduct.[187] Therefore, as a matter of principle, and to enhance deterrence, Article 81(1) EC must prohibit unsuccessful attempts: a presumption of allocative inefficiency based on intent cannot be rebutted by measurement *ex post*.[188]

(ii) Ex ante If the collusive conduct has yet to be implemented, there is a question of whether economic prediction can be used to rebut the legal presumption of allocative inefficiency based on the predictive quality of intent. The economic predictive tools identify conditions in which it is possible to contrive scarcity.[189] If the economic conditions do not obtain, *an intention to cause* allocative inefficiency *cannot cause* allocative inefficiency.[190] This suggests that the presumption *ought to be* rebuttable.[191]

However, such a position assumes that economic prediction offers more reliable prediction of outcome than the epistemic quality of intent. Yet, a claim can be made that intent provides better prediction than economics.[192] Korah and O'Sullivan seem to suggest this when writing that because the undertaking 'is risking its profits and livelihood when it makes decisions... It probably understands the product and market better than any official or a court'.[193] Does this not accept that party intent and not economics gives the best indication of the likelihood of success? Whether or not the presumption of allocative inefficiency based on the predictive quality of intent is rebuttable or conclusive turns on where intent and economic evidence lie in the hierarchy of evidence used to predict

[185] Posner (1976: 11). [186] Posner (1976: 12). Also Bork (1993: 112–113, 123).

[187] Rent seeking and its costs are more fully explained in Chapter 6: Section II.B

[188] That US antitrust prohibits unsuccessful attempts to restrict competition is established in *United States v Socony-Vacuum Oil Co* 310 US 150, 224, footnote 59 (1940).

[189] See Section III.B above. [190] Guerrin and Kyriazis (1993: 798).

[191] Hawk (1995: 977 n 6), Posner (1976: 25, 47), Bork (1993: 116), Comments (1952), and Brennan (2001: 1091–1093).

[192] In *Brooke Group Ltd v Brown & Williamson Tobacco Corporation* 509 US 209, 249, 253 (1993), dissenting opinion, Justice Stevens considered intent more probative than economic theory.

[193] Korah and O'Sullivan (2002: 29). Also Posner (2003: 96), Neven *et al.* (1998: 21), and Ehlermann and Drijber (1996: 381).

outcomes.[194] Coase gives a general warning against a reliance on economics unless it is shown to provide greater insight than other approaches.[195] Whether and when the Court does, or will, attach more probative value to the external assessment of economics than to the internal assessment of the undertakings, expressed by intent, remains to be seen.

IV. Conclusion

This chapter has been concerned to identify the meaning attached to the term 'restriction of competition' in Article 81(1) EC. The view that the term serves merely to establish jurisdiction and so requires little substantive assessment has been considered. This view can be seen to reflect past practice, but it is not the approach to be taken in the future. Instead, the term 'restriction of competition' in Article 81(1) EC must serve a substantive function. It is argued that the obligation appropriate to impose on entities engaged in economic activity (undertakings) is not to collude to cause economic inefficiency. It is argued that allocative efficiency is either measured, predicted, or presumed to exist in all cases in which a restriction of competition is said to exist.[196] This is illustrated in Figure 5.2.

Whether the obligation not to collude to cause allocative inefficiency has been breached can be demonstrated *ex post* by measurement. Care must be taken to measure consequences for the market rather than consequences for a particular firm (when the two are properly distinct). Measurement of market output and price can only be carried out in an *ex post* analysis, ie, after the claimed anti-competitive consequences have occurred. A system of *ex ante* control requires tools to predict whether anti-competitive outcomes are more likely than not. Whether the obligation not to collude to cause allocative inefficiency will be breached can be demonstrated *ex ante* with economic predictive tools; many debates in competition law centre on whether the tools available enable accurate prediction of the anti-competitive consequences we seek to guard against. The distinction between *ex post* and

[194] Compare Posner (1976: 41), Bork (1993: 123), Williamson (1987: 297–298), and Anlauf (2000: 506). This argument is made more fully in Odudu (2002b: 484–490).

[195] Coase (1994). Patterson (2000: particularly at 435 n 31) argues that economic tools are not the only way to determine the ability to contrive scarcity of output.

[196] Muris (1998: 773, 790) advocates the view that there are many ways in which allocative inefficiency can be identified.

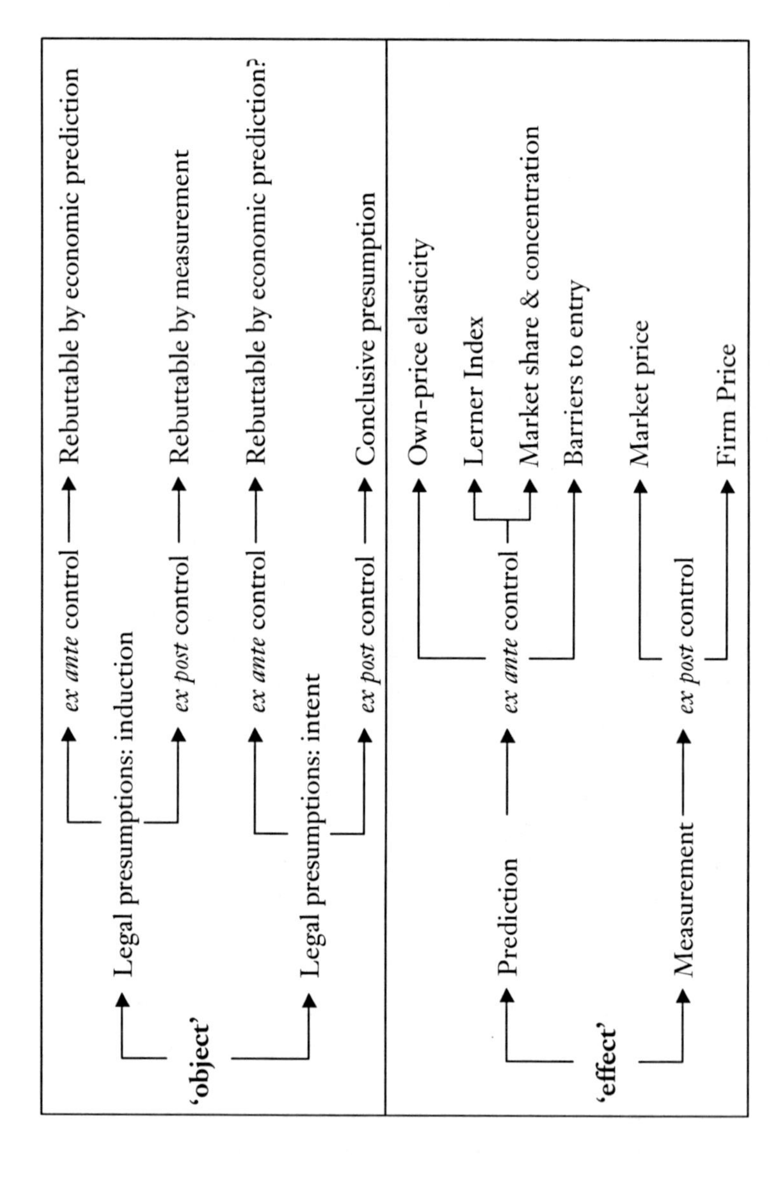

Figure 5.2 Demonstrating Allocative Inefficiency in Article 81(1) EC.

ex ante Article 81(1) EC intervention draws attention to the different types of evidence required in the different contexts. Allocative efficiency demonstrated by measurement or prediction is said to have the *effect* of restricting competition. There are also legal presumptions, developed from experience or based on party intent, that certain collusive conduct will result in allocative inefficiency. Presumptions based on experience ought to be rebuttable, and seem to be so. However, presumptions based on party intent seem to be irrebuttable. In this sense, an attempt to restrict competition is treated as if it restricts competition, which enhances the deterrent effect of the competition rules. Allocative inefficiency established by legal presumption is said to have the *object* of restricting competition. The extent to which it is necessary to show the *effect* of restricting competition is determined by the scope and content of what is said to have the *object* of restricting competition; the most intensive intervention centres on cases involving collusion legally presumed to result in allocative inefficiency.[197] Great attention should be placed on the content of the legal presumptions and how they are established.

[197] White Paper on Modernisation of the Rules Implementing Articles 85 and 86 of the EC Treaty Commission Programme No 99/027 [1999] OJ C 132/1, executive summary para 13, third recital of Council Regulation (EC) No 1/2003 [2003] OJ L 1/1, and Venit (2003: 546).

6

Article 81(3) EC as a productive efficiency enquiry

I. Introduction

The previous chapter attempted to show that a restriction of competition within Article 81(1) EC means allocative inefficiency and is proven by measurement, economic prediction or legal presumption. Though Article 81(2) EC makes collusion caught by the prohibition automatically void, it is recognized that an exclusively allocative efficiency concern is too narrow a conception of the competition law problem.[1] Attention is also paid to whether conduct increases, is neutral, or reduces productive efficiency, which has a greater impact on living standards.[2] Concluding his inaugural lecture, Vickers expressed the view that '[r]easoning about competition problems requires better theoretical understanding . . . of how competition . . . works . . . In particular it requires frameworks that explicitly address effects on productive efficiency.'[3] This chapter argues that Article 81(3) EC provides a framework in which to assess an undertaking's claim that the allocative inefficiency consequences of their collusion are outweighed by productive efficiency gains.[4]

Productive efficiency is the relationship between the output of goods and the input of resources used to make them and has two

[1] Chapter 2: Section II.B On Article 81(2) EC: Goyder (2003: 138–139).

[2] Kendrick (1977: 1), Schumpeter (1975: 83), Nelson and Winter (1982: 114), Romer (1986: 1003), Romer (1996: 204), Ahdar (2002: 344), Williamson (1968a: 22–23), and Vickers (1995: 7).

[3] Vickers (1995: 18). Compare with Bork (1993: 104).

[4] Guidelines on the Application of Article 81(3) of the Treaty [2004] OJ C 101/97, para 33, Guidelines on the Applicability of Article 81 of the EC Treaty to Horizontal Cooperation Agreements [2001] OJ C 3/2, para 4, Guidelines on Vertical Restraints [2000] OJ C 291/1, para 5, and Case T-112/99 *Métropole Télévision (M6) v Commission* [2001] ECR II-2459, para 77.

components: technology and technical efficiency.[5] Technology describes the methods by which inputs are transformed into outputs. Technical efficiency describes being able to maximize outputs given input quantities, or minimize inputs given output quantities, using a given technology.[6] The rate at which the underlying technology changes is also included in the conception of productive efficiency advanced here, recognizing that 'progress to a great extent involves the creation of a better article, rather than the production of an old article for a cheaper price'.[7] Thus Porter argues that all 'practices scrutinized in antitrust should be subjected to the following question: How will they affect productivity growth?'[8]

Section II examines reasons to suppose that undertakings able to profitably contrive scarcity of output are *less* productively efficient than are undertakings without market power. Conversely, Section III examines reasons to suppose that undertakings able to profitably contrive scarcity of output are *more* productively efficient than are undertakings without market power. Both allocative and productive efficiencies are relevant to living standards and it may be necessary to forgo allocative efficiency to obtain greater productive efficiencies. Section IV considers how Article 81(3) EC structures the productive efficiency enquiry and the trade-off between productive and allocative inefficiency. Section V concludes the chapter.

II. The productive *in*efficiency of firms with market power

Whilst Section III below considers how productive efficiency can act as a counterbalance to allocative *in*efficiency, it is important to consider arguments that collusion between undertakings with the aim or consequence of reduced output and increased price is also likely to result in productive *in*efficiency.[9] The first source of productive inefficiency is termed x-inefficiency, the second source is rent-seeking behaviour. These add to, rather than counterbalance, the allocative inefficiency concern. It is

[5] Kendrick (1977: 14). [6] Grosskopf (1993: 160) and Farrell (1957: 254).

[7] Boulding (1945: 539). Also Schumpeter (1975: 83), Nelson and Winter (1982: 121), Aghion and Howitt (1992: 349), Economist (2002), Porter (2001: 922–924), Viscusi *et al.* (2000: 88–89), and Jones (1998: 72). [8] Porter (2001: 932). Also Stelzer (2002).

[9] Tirole (1988: 75).

acknowledged that these 'cost distortions and rent-seeking behaviours have not yet been mastered by economists'.[10]

A. *X-inefficiency*

Collusion between undertakings that aims at, or results in, reduced output and increased price may also result in productive inefficiency if undertakings able to profit from contrived scarcity of output are less *willing* or less *able* to minimize costs. In the absence of market power, undertakings are thought of as striving to gain an advantage over their competitors by reducing the cost of production and thus increasing profits.[11] Undertakings with market power, ie, those able to profit from contrived scarcity of output, do not strive to reduce costs, hence consume more resources to produce their output than do firms without market power.[12] The *unwillingness* to minimize costs arises because though undertakings employing human capital purchase a person's *time*, production requires a person's *effort*.[13] Employees enjoy a degree of discretion as to how much effort to exchange for their remuneration.[14] The level of effort employees make varies with the competitiveness of the market; employees face external pressure to increase effort and thus minimize cost because 'if one firm can lower its price then all firms must follow suit or be eliminated from the industry. In the long run firms survive by keeping costs as low as possible.'[15] Employees in undertakings that possess market power face insufficient pressure to increase effort. Because the undertaking can produce inferior goods without fear of losing sales, the employees do not fear for their jobs.[16]

The idea that undertakings are unwilling to reduce costs is objected to because it conflicts with the idea that undertakings are profit maximizers.[17] An alternative explanation for the existence of x-efficiency, retaining the assumption of profit maximization, is that firms with market power are less *able* to reduce cost than those without market power.[18] Undertakings in

[10] Tirole (1988: 78). [11] Lutz (1989: 161).

[12] Frantz (1988: 54, 64–65), Scherer (1987: 1004–1010) and Schaub (1998: 120).

[13] Frantz (1988: 74) and Tirole (1988: 75).

[14] Leibenstein (1966: 407) and Frantz (1988: 55, 74–89).

[15] Frantz (1988: 97–98, also 57–60).

[16] Leibenstein (1966: 408–409), Frantz (1988: 95), Hicks (1935: 8), Hildebrand (1998: 144), Faull and Nikpay (1999: para 1.65–1.66), Williamson (1968a: 31), and Economist (2000a).

[17] Stigler (1976). X-efficiency theory is defended against various criticisms in Frantz (1992), Frantz (1988: 5–7, 21, 46–53, 183–199) and Leibenstein (1988: xvii).

[18] It may be that collusion simply enables the less efficient to survive: Motta (2004: 51).

competitive industries compare their performance with that of others, making inferences about how much effort employees make.[19] The more market power undertakings obtain, the more difficult it is to find like undertakings with which to make a comparison and the more difficult it is to know whether internal costs are as low as they can be.[20]

B. *Rent seeking*

A further source of productive inefficiency for undertakings able to profitably contrive scarcity of output is the cost of creating and protecting market power—rent-seeking.[21] As expressed by Tirole, '[f]irms will tend to spend money and exert effort to acquire [market power]; once installed in that position, they will tend to keep on spending money and exerting effort to maintain it'.[22] Posner argues that all super-normal profit will be dissipated in the effort to achieve and protect market power.[23] Thus, the cost of market power, in addition to allocative inefficiency, includes the cost of achieving and maintaining market power—all the super-normal profits.[24]

Whilst recognizing that rent seeking wastes some resources, it is unlikely that *all* super-normal profit will be dissipated in the effort to earn it and that *all* the effort expended is socially wasteful.[25] Further, although Posner is correct in including the cost of those who exert effort but are unsuccessful in obtaining market power, Williamson notes that those with a low possibility of securing the rent are unlikely to invest any resources in seeking it.[26]

III. The productive efficiency of firms with market power

In Section II above, reasons suggesting that undertakings colluding with the object or effect of restricting competition not only cause allocative inefficiency but also cause productive inefficiency are discussed. However,

[19] Tirole (1988: 41–42, 76) and Porter (2001: 928–929).

[20] Vickers (1995: 8–12) and Carlton and Perloff (2000: 93).

[21] Frantz (1988: 28–34, 184–187) and Frantz (1992: 436–437) considers rent seeking and x-efficiency as alternative explanations.

[22] Tirole (1988: 76).

[23] Posner (1975: 809).

[24] Posner (1975: 821–822). Compare with Bork (1993: 112–113). There is also the cost of competition law enforcement trying to prevent a supernormal profit being earned: Posner (1975: 807–812), Faull and Nikpay (1999: para 1.67–1.68), Williamson (1968a: 24), and Nealis (2000: 367–368).

[25] Fisher (1985), Tirole (1988: 76–78), and Williamson (1977).

[26] Williamson (1977: 716–721).

collusion between undertakings may occur precisely to minimize production costs.[27] A profit-maximizing undertaking must decide whether to perform a particular task itself, or engage another undertaking to perform the task. When the task is more effectively performed externally, including the costs associated with finding and monitoring external performance, performing the task externally reduces costs.[28] When undertakings collude it may well indicate that the costs of non-collusive production are higher than the costs of collusive production.[29] There are a number of reasons to believe that undertakings able to profitably contrive scarcity of output are able to achieve productive efficiencies, and that market power perhaps makes them more able. Colluding undertakings may be able to exercize countervailing power, exploit economies of scale and scope, and make technological progress.[30] These sources of productive efficiency are examined below.

A. *Countervailing power*

In a competitive market, undertakings are prevented from obtaining and exploiting market power by *competitor*-undertakings that offer, or threaten to offer, a better bargain.[31] Galbraith argues that undertakings can also be prevented from exploiting market power by *customer*-undertakings with market power. Customer-undertakings with market power can effectively 'force lower prices' from the holder of the original market power.[32] Dobson and Waterson examined the idea that countervailing power developed by a retailer can be deployed against a manufacturer with market power to obtain goods or inputs at a lower price, finding that the more power the retailer develops, the greater the discount extracted.[33] The claim is that inequality of bargaining power is necessary for the exercize of market power: competition ensures that both sides are weak; countervailing power

[27] Williamson (1974: 1450–1463), Williamson (1977: 723–726), and Winsten and Hall (1961: 259).

[28] Coase (1937) and Williamson (1975).

[29] eg, fifth recital of Commission Regulation No 1984/83 [1983] OJ L 173/5, Green Paper on Vertical Restraints in EC Competition Policy Com (96) 721 Final [1997] 4 CMLR 519, para 125, Blair and Kaserman (1985: 409, 413), and Dobson and Waterson (1996: 18).

[30] Technological progress can be treated as dynamic efficiency, distinct from rather than a constituent of productive efficiency. Here the two are not distinguished.

[31] Galbraith (1957: 112).

[32] Galbraith (1957: 118, 115, 123) and Hunter (1958: 103). Compare with Guidelines on the Assessment of Horizontal Mergers under the Council Regulation on the Control of Concentrations between Undertakings [2004] OJ C 31/5, para 64–67.

[33] Dobson and Waterson (1997: 420–426).

ensures that both sides are strong.[34] The importance of the claim is that, if true, 'it relieves the government of its obligation—imposed by the now antiquated antitrust laws—to launch any frontal attack on concentrated economic power'.[35]

Though customer-undertakings are not *exploited*, the crucial question is why customer-undertakings do not *exploit* their power against the end consumer, ie, why are savings passed on. Galbraith considered that those developing countervailing power do so 'by proxy, on behalf of the individual consumer'.[36] Yet, as Hunter notes, 'one can hardly base an important welfare theorem on the assumption that oligopsonists are altruistic'.[37] What ensures that those developing countervailing power do so *on behalf of the individual consumer* is not altruism, but intense competition in the market in which countervailing power exists. Otherwise countervailing power simply leads to higher consumer prices.[38] We must therefore be cautious when parties falling within the Article 81(1) EC prohibition claim they have done so to develop countervailing power. If the conditions for countervailing power being exercized on behalf of consumers are satisfied, it is difficult to see how the Article 81(1) EC conditions are infringed.[39]

B. *Economies of scale and scope*

Exploiting economies of scale can increase productive efficiency.[40] Economies of scale exist if the average cost of production falls as the quantity of goods produced increases.[41] Achieving economies of scale is one of the underlying aims of the European Community.[42] Three reasons are offered for the existence of economies of scale. *Indivisibility*: a minimum number of inputs are required to be in business, regardless of how much is produced. This same number of inputs can be used to produce various output levels,

[34] Galbraith (1957: 111–112). [35] Adams (1953: 472).

[36] Galbraith (1957: 126, also 117).

[37] Hunter (1958: 101, 100). Also Stigler (1954: 9–11), Galbraith (1959), and Hunter (1959).

[38] Dobson and Waterson (1997: 420–426), Hunter (1958: 99–101), Adams (1953: 475–477), Galbraith (1954: 3–4), and Galbraith (1957: 117–118).

[39] Compare with Gyselen (2002: 188–191). Countervailing power is not durable since the party holding the original market power can avoid it by vertical agreements or vertical integration: Adams (1953: 472–474), and Dobson and Waterson (1997: 426–428).

[40] Kendrick (1977: 69), Scherer (1987: 1002–1003), and Hunter (1958: 90–91). Diwan (1966: 451–453) attempts to estimate the increase in output derived from scale economies.

[41] Carlton and Perloff (2000: 35), Silberston (1972: 369), and Van den Bergh and Camesasca (2001: 182–184). [42] Wesseling (2000: 10–11) and Chapter 2: Section III.

so the cost is spread over a greater number of units as output increases.[43] *Better technology*: scale also produces economies because technology that is more efficient can be used as the amount to be produced justifies its employment.[44] Silberston writes, '[t]he higher the rate of output . . . the more scope there is likely to be for mechanised processes which would be uneconomical at smaller rates of output'.[45] *Specialization*: economies of scale also exist because higher levels of output allow inputs to be used for specialized tasks.[46] Absent specialization, time is lost 'passing from one species of work to another'.[47] Further, workers get better at doing a particular task with experience.[48]

In addition to scale economies, a firm may be able to exploit economies of scope. These exist when it is cheaper to produce two or more products together than it is to produce them separately.[49] Savings arise when production of distinct products requires sharable or common inputs, and the cost of providing the input jointly is less than providing it separately.[50] It is important to note that whilst economies of scale and scope may be created in certain dimensions, these savings may be outweighed by *dis*economies in other dimensions and our concern must be with the overall benefit.[51]

C. *Technological progress*

Living standards increase not simply by having more and thus cheaper goods, but by having 'new consumers' goods, the new methods of production or transportation, the new markets, the new forms of industrial organization that capitalist enterprise creates'.[52] Undertakings create new ideas, new production methods, and new goods through research and

[43] Begg *et al.* (2000: 112–113) and Silberston (1972: 374). The extent to which indivisibilities account for economies of scale is critiqued in Chamberlin (1948: 244–246), McLeod and Hahn (1949), and Chamberlin (1949).

[44] Chamberlin (1948: 236, 242–244) and Begg *et al.* (2000: 113).

[45] Silberston (1972: 371, also 369).

[46] Carlton and Perloff (2000: 37), Silberston (1972: 374–375), Chamberlin (1948: 236), and Begg *et al.* (2000: 113).

[47] Chamberlin (1948: 240 citing Adam Smith).

[48] Begg *et al.* (2000: 514), Scherer and Ross (1990: 98, 372–373), and Carlton and Perloff (2000: 265).

[49] Carlton and Perloff (2000: 44), Panzar and Willig (1981: 268), Dobson and Waterson (1996: 26), and Van den Bergh and Camesasca (2001: 185–186).

[50] Panzar and Willig (1981) and Carlton and Perloff (2000: 45).

[51] Carlton and Perloff (2000: 36–38), Chamberlin (1948: 247–255), and Van den Bergh and Camesasca (2001: 184–185).

[52] Schumpeter (1975: 83).

development activities.[53] The quantity and quality of current resources devoted to research and development depends on how profitable the future innovation will be for the undertaking; the ability to profitably contrive scarcity of output affects the profitability of innovation.[54] Market power, and the allocative inefficiency that this entails, is seen by some as essential to technological progress and thus productive efficiency. We are thus asked to forgo current consumption and allocative efficiency to incentivize or enable investments resulting in technological progress, as this creates more to allocate in the future.[55] The point cannot be made more forcefully than by repeating Schumpeter's words:

> A system—any system, economic or other—that at *every* given point of time fully utilizes its possibilities to the best advantage may yet be inferior to a system that does so at *no* given point in time, because the latter's failure to do so may be a condition for the level or speed of long-run performance.[56]

Two alternative reasons for the dependency of technological progress on market power (and thus allocative inefficiency) are offered: first, it is thought that only those making super-normal profits have the *ability* to invest in innovation; second, it is thought that undertakings do not have the *incentive* to invest in innovation unless they are rewarded for their investment by super-normal profit.[57] These reasons are assessed in turn.

1. *Ability to innovate*

Market power is first said to give undertakings the ability to invest in research and development. The idea is that 'only bigness can provide the sizable funds necessary for technological experimentation and innovation'.[58] Undertakings invest supra-normal profits in innovation.[59]

Adams rejects the idea because history shows that undertakings with market power have not been leaders in innovation.[60] Rather, he suggests that small undertakings are at least as innovative, but less successful in

[53] Kendrick (1977: 67–69), Romer (1986: 1007), Jones (1998: 84, 89–90), Carlton and Perloff (2000: 506), and Aghion and Howitt (1992: 324).

[54] Viscusi *et al.* (2000: 93), Economist (2002), Cornish (1996: 13–14, 30), Bainbridge (1999: 322), and Dutton (1984: 3, 17–23).

[55] Romer (1986: 1015, 1019), Nelson and Winter (1982: 114, 116), Aghion and Howitt (1992: 331), Carlton and Perloff (2000: 505), Williamson (1969b: 105–106), Williamson (1977: 712), and Hildebrand (2002a: 8).

[56] Schumpeter (1975: 83, emphasis in original).

[57] Scherer (1987: 1010–1018) and Scherer (1992).

[58] Adams (1953: 477).

[59] Nelson and Winter (1982: 115–116, 130) and Stelzer (2002: 27–28).

[60] Adams (1953: 478–480).

achieving government favour and protection.[61] Though debatable whether undertakings with market power are more innovative, the weight of opinion bends towards the view that the presence or absence of market power has little bearing on the *ability* to innovate.[62] Scherer sums this up, writing:

> The only simple conclusion stemming from . . . theoretical research stimulated by Schumpeter's original conjectures is that the links between market structure, innovation, and economic welfare are extremely complex. . . . where technological progress is most vigorous—with loose oligopoly, tight oligopoly, or pure monopoly, with or without impeded entry—remains unclear a priori, varying from case to case.[63]

2. *Incentive to innovate*

A more convincing rationale for allowing firms to exercize market power is that this provides an incentive to invest in innovation: '[o]nly monopoly earnings can provide the bait that lures the capital onto untried trails'.[64] This incentive is required because ideas face non-rivalry in consumption.[65] Once produced, an idea can be used by an infinite number of people indefinitely without deteriorating.[66] Since the cost of producing the idea remains fixed, regardless of the numbers consuming it, there is a zero marginal cost in allowing additional consumption.[67] Article 81(1) EC strives towards allocative efficiency; this is achieved when firms price at marginal cost.[68] If marginal cost pricing were enforced in relation to ideas, the cost of producing the idea could never be recovered and there would be no incentive to produce it.[69] Undertakings will not invest in producing ideas unless there is some reward.[70] Allowing successful innovator-undertakings to contrive scarcity of output and thus to command an increased price provides an incentive to innovate.[71]

[61] Adams (1953: 488–491).

[62] Carlton and Perloff (2000: 533), Faull and Nikpay (1999: para 1.122–1.123), Fels and Edwards (1998: 59–63), Rapp (1995: 28–30), Scherer (1992: 1425–1430), Van den Bergh and Camesasca (2001: 36–37), Viscusi *et al.* (2000: 93–94), Williamson (1968a: 29–31), and Williamson (1969b: 115–116).

[63] Scherer (1992: 1421).

[64] Adams (1953: 478).

[65] See Chapter 3: Section II.C.2, accompanying notes 111–112.

[66] Jones (1998: 73), Romer (1996: 204), and Carlton and Perloff (2000: 505).

[67] Romer (1986: 1007).

[68] See Chapter 5: Section III.

[69] Jones (1998: 78–79) and Romer (1986: 1003, 1020).

[70] Nelson and Winter (1982: 117), Boulding (1945: 527), Jones (1998: 74–76, 79), Economist (2000c), and Carlton and Perloff (2000: 505–507). Carlton and Perloff (2000: 513–532) suggest various methods used to encourage innovative activity.

[71] Nelson and Winter (1982: 115), Carlton and Perloff (2000: 501, 510–511), Bork (1993: 134–135), Economist (2001) and Economist (2000b: 130). Mario Monti, then Commissioner for Competition, is reported as being sceptical of this argument: Economist (2000e).

The ability to profitably command a higher price from contrived scarcity is destroyed by the next innovation that comes along; market power 'based on superior efficiency . . . does comparatively little harm so long as it is assured that it will disappear as soon as anyone else becomes more efficient in providing satisfaction to the consumers'.[72] However, the fragility of this market power creates its own competition law concerns. Those currently enjoying the fruits of innovation have an incentive to preserve the *status quo*, not by further innovation, but by stifling the attempts of others to innovate.[73] Vigilance is necessary to identify attempts to protect a position other than by innovation.[74]

IV. Article 81(3) as a productive efficiency enquiry

Sections II and III above consider how undertakings able to profitably contrive scarcity of output affect productive efficiency. It is clear that collusion causing allocative inefficiency has consequences in the productive efficiency dimension. Further, it is clear that both efficiencies affect living standards, so that an exclusive concern with allocative inefficiency is unwarranted. Chapter 5 sought to show that allocative efficiency is addressed in an Article 81(1) EC enquiry. It remains 'controversial what exactly remains to be considered under Article 81(3)'.[75] By examining its conditions and how they are satisfied it is first argued that Article 81(3) EC can be seen to involve a productive efficiency enquiry. It is then considered how Article 81(3) EC structures an enquiry into whether productive efficiency gains outweigh the allocative efficiency loss. In the end, it is possible to see the Article 81(3) EC elements as ensuring that efficiency as a whole is enhanced and that consumers benefit from the enhanced efficiency.

A. *The need for productive efficiency*

The first positive condition contained in Article 81(3) EC states that Article 81(1) EC may be declared inapplicable if the agreement or conduct

[72] Hayek (1980: 105). Also Schumpeter (1975: 83), Aghion and Howitt (1992: 324, 331, 335–336), and Economist (2000d).

[73] Economist (1999), Economist (2000c), Elhauge (2002), and Economist (2000b: 130).

[74] Beutel (2002).

[75] Wesseling (2001: 369, also 362), Whish and Sufrin (2000: 146–149), Whish (2000a: 75–78), Monopolkommission (2000: 27), and Panel Discussion (1998: 480, 488, 491). Compare with the approach taken in Gyselen (2002: 188–191).

'contributes to improving the production or distribution of goods or to promoting technical or economic progress'. This is and ought to be understood as a productive efficiency requirement.[76] '[L]ower costs, new products or services or products or services of a better quality' satisfied the then Commissioner Monti that the condition was fulfilled.[77] Gyselen considers the condition satisfied by 'a new or upgraded technology, product or service... cost savings and lower prices... a better pre- or after-sales service'.[78] Faull and Nikpay list cost reductions and productivity increases, improvements in quality and choice, and improvements in specific technologies as factors to be considered under the first positive condition.[79] Bellamy and Child consider the condition satisfied when collusion is 'likely to lead to a better use of resources, such as rationalisation of production, the achievement of economies of scale, or the faster more effective development of new products'.[80] Continuity of supply, new products more quickly and at lower cost, a product of higher quality, and the entry of a new competitor or a wider range of choice of supplier have all been considered benefits.[81] The Commission and Court simply refer to these factors as 'beneficial economic effects'.[82] A comparison with the factors considered in Sections II and III above makes it apparent that these 'beneficial economic effects' are productive efficiency gains.[83]

That Article 81 EC ought to be applied in a manner that preserves both productive efficiencies and the incentives to generate them is increasingly recognized.[84] It is recognized that, whilst causing allocative inefficiency,

[76] Guidelines on the Application of Article 81(3) of the Treaty [2004] OJ C 101/97, para 50, Neven (1998: 112), and Wolf (1998: 131). American Bar Association (Antitrust Section) and Hartley (chairman) (1999: 116–119) describes one aspect of rule of reason analysis as a productive efficiency enquiry. Compare with Venit (2003: 561 note 56, 576–577), Matte QC (1998: 101), and Goldman QC and Barutciski (1998: 406–408). On productive efficiencies in EC merger analysis: Levy (2003: para 15.01–15.02).

[77] Monti (2000: 6).

[78] Gyselen (2002: 185).

[79] Faull and Nikpay (1999: para 2.127–2.153, 2.160).

[80] Rose (1993: para 3-022, also para 3-040).

[81] Rose (1993: para 3-039), Ritter *et al.* (2000: 124–125), and Van den Bergh and Camesasca (2001: 207–209).

[82] Guidelines on the Applicability of Article 81 of the EC Treaty to Horizontal Cooperation Agreements [2001] OJ C 3/2, para 32 and Roth QC (2001: para 3-026).

[83] Guidelines on the Application of Article 81(3) of the Treaty [2004] OJ C 101/97, para 59–72, and Guidelines on the Applicability of Article 81 of the EC Treaty to Horizontal Cooperation Agreements [2001] OJ C 3/2, para 102, 132, 151. Compare with Art 2(1)(b) ECMR, Guidelines on the Assessment of Horizontal Mergers under the Council Regulation on the Control of Concentrations between Undertakings [2004] OJ C 31/5, para 76–88, Levy (2003: para 15.02(2)), and ICN Merger Analytical Framework Subgroup (2004: para 34–64).

[84] Guidelines on the Application of Article 81(3) of the Treaty [2004] OJ C 101/97, para 50, 80–81, Case C-7/1997 *Oscar Bronner GmbH & Co. KG v Mediaprint Zeitungs- und Zeitschriftenverlag*

collusion resulting in the ability to profitably contrive scarcity of output can encourage investment:

> [f]ree from intrabrand price competition, the intermediary has sufficient *incentive to invest* in plant, equipment, and knowledge development, which can add value-enhancing services to the manufacturer's goods and services. . . . *The assurance of sufficient margins* via intrabrand price protection *is the catalyst to such value-enhancing investment*.[85]

The Commission recognizes that collusion 'can be a means to share risk, save costs, pool know-how and launch innovation faster'.[86] It is acknowledged that collusion may promote the 'cross fertilisation of ideas and experience, thus resulting in improved or new products and technologies being developed more rapidly than would otherwise be the case'.[87] It is accepted that collusion may be necessary to obtain economies of scale.[88]

Though productive efficiencies are recognized as relevant there is a question of whether they are relevant under Article 81(1) EC or under Article 81(3) EC. Productive efficiencies have in the past been considered as part of an Article 81(1) EC analysis.[89] Article 81(3) EC enables a more rigorous assessment of whether Article 81(1) EC has in fact been infringed,

GmbH & Co. KG [1998] ECR I-7791, AG Opinion para 57, and Korah (2002: 802–807). Productive efficiencies are explicitly part of EC merger analysis: Levy (2003: para 15.02).

[85] Zerrillo *et al.* (1997: 711, emphasis added). Also Sixth recital of Commission Regulation No 1983/83 [1983] OJ L 173/1, Blair and Kaserman (1985: 369–370, 409–411), Tirole (1988: 182–185), Van den Bergh and Camesasca (2001: 225–227), Liebeler (1982: 5), Griffiths (2000: 245), Whish (2001: 542), Carlin (1996: 284), and Korah and O'Sullivan (2002: 26).

[86] Guidelines on the Applicability of Article 81 of the EC Treaty to Horizontal Cooperation Agreements [2001] OJ C 3/2, para 3.

[87] Guidelines on the Applicability of Article 81 of the EC Treaty to Horizontal Cooperation Agreements [2001] OJ C 3/2, para 68.

[88] Fifth recital of Commission Regulation No 1983/83 [1983] OJ L 173/1, Guidelines on Vertical Restraints [2000] OJ C 291/1, para 174, Maison Jallatte [1966] JO 37/66 [1966] CMLR D1, D4, Green Paper on Vertical Restraints in EC Competition Policy Com (96) 721 Final [1997] 4 CMLR 519, para 122, and Blair and Kaserman (1985: 369).

[89] Case 161/84 *Pronuptia de Paris GmbH v Pronuptia de Paris Irmgard Schillgalis* [1986] ECR 353, para 15, Case 48/72 *Brasserie de Haecht SA v Wilkin* [1973] ECR 77, para 2–5, Case 59/77 *De Bloos v Bouyer* [1977] ECR 2359, para 11, Hawk (1988: 64–65, 69–70), Hawk (1995: 974, 977–978, 980, 982–983, 986–988), Wesseling (2000: 88–93, 103–105), Faull and Nikpay (1999: para 2.57, 7.40), Green (1988: 199–201), Deckert (2000: 178–179), Korah (2000: 237), Korah (1994), Siragusa (1998), Bright (1995: 515), Bright (1996), Forrester and Norall (1984: 23–24, 38–40), Hildebrand (1998: 188, 219, 235), Goyder (1998: 34–54), White Paper on Modernisation of the Rules Implementing Articles 85 and 86 of the EC Treaty Commission Programme No 99/027 [1999] OJ C 132/1, para 49, Gyselen (2002: 188–191), Korah (1993c: 65–66), Whish (1993: 207), Hawk (1988: 67–69, 70, 72), Houtte (1982), and Black (1997). Compare with Gonzalez-Diaz (1996) and Hurley (1987).

on the basis that when 'faced with a strong efficiency explanation for a course of conduct, a court is more likely to reject the anticompetitive explanation than in its absence'.[90]

Three problems with treating productive efficiencies as relevant under Article 81(1) EC are encountered. First, whether productive efficiencies are considered has depended on how the allocative inefficiency is caused. Productive efficiencies are not considered under Article 81(1) EC when allocative inefficiency is caused by price fixing or market sharing agreements.[91] The Court considers that though such restrictions may be necessary to achieve productive efficiency, this fact 'is relevant only to an examination of the agreement in the light of the conditions laid down in Article [81(3)]'.[92] Second, the rationale for conducting a balance of allocative and productive efficiency within Article 81(1) EC was that it avoided the then procedural implications of a division of competences brought about by the direct effectiveness of Article 81(1) EC, and the exclusive role of the Commission in applying Article 81(3) EC.[93] National courts under a duty to apply Article 81(1) EC were unable to apply Article 81(3) EC.[94] Commenting on this situation, Advocate General Lagrange felt that:

> it would be contrary to the most elementary considerations of justice to allow the application of the prohibition contained in Article [81(1)], with the sanction of nullity which attaches to it, together with all the other consequences which the municipal courts could have, and perhaps should apply, without allowing the enterprises concerned to avail themselves of the provisions of Article [81(3)].[95]

Wesseling expresses the view that '[i]nterpretations of Community substantive law generally, and of EC antitrust law in particular, are interrelated and *dependent* upon institutional, *procedural*, political and economic aspects of the system'.[96] Considering productive efficiencies as part of the Article 81(1) EC analysis can be seen as one of the many means developed to circumvent the problem created by the Commission's monopoly over the

[90] Kattan (1996: 615, also 617 text accompanying notes 18–20), Gyselen (2002: 189), Williamson (1977: 728), Van den Bergh and Camesasca (2001: 163–164), Camesasca (1999: 18, 23, 25–26), and Bork (1993: 221–222).

[91] Case 161/84 *Pronuptia de Paris GmbH v Pronuptia de Paris Irmgard Schillgalis* [1986] ECR 353, para 24–25.

[92] Case 161/84 *Pronuptia de Paris GmbH v Pronuptia de Paris Irmgard Schillgalis* [1986] ECR 353, para 24.

[93] Article 9(1) of Regulation 17. First Regulation Implementing Articles 81 and 82 of the Treaty [1962] OJ Special Edition 204/62.

[94] Whish (2001: 262–267).

[95] Case 13/61 *Kledingverkoopbedrijf de Geus En Untdenbogerd v Robert Bosch GmbH* [1962] ECR 45, 66.

[96] Wesseling (2000: 4, emphasis added, also 19, 28). *Contra* Bright (1996: 544–545).

exercise of Article 81(3) EC. However, Council Regulation 1/2003 now rectifies the procedural anomaly.[97] Finally, the Commission has always rejected the idea that *all* the economic analysis should take place within Article 81(1) EC. Consideration of productive efficiencies in Article 81(1) EC would leave something other than efficiency to be considered in Article 81(3) EC, which is denied.[98] Instead, the purpose of Article 81(3) EC 'is to provide a legal framework for the economic assessment of restrictive practices and *not to allow application of the competition rules to be set aside because of political considerations*'.[99]

In *European Night Services*, the CFI considered whether productive efficiencies are relevant and whether they were relevant under Article 81(1) EC or Article 81(3) EC. The undertakings argued that economies of scale and incentives to take financial risk were relevant, and relevant under Article 81(1) EC.[100] The Commission recognized the productive efficiency claims as relevant, but relevant under Article 81(3) EC.[101] Though hardly unequivocal, the CFI endorses the Commission view.[102] The question was again considered in *Métropole*.[103] It was predicted that allocative inefficiencies would result from an exclusive purchase/distribution agreement.[104] The undertakings argued that productive efficiencies also existed, and that these should also be weighed under Article 81(1) EC.[105] The CFI rejected

[97] Council Regulation (EC) No 1/2003 [2003] OJ L 1/1.

[98] White Paper on Modernisation of the Rules Implementing Articles 85 and 86 of the EC Treaty Commission Programme No 99/027 [1999] OJ C 132/1, para 56, Hawk (1988: 69), Siragusa (1998: 665–666), Monopolkommission (2000: 38), Mavroidis and Neven (2001: 157–159), Furse (1996: 257–258), and Pitofsky (1995: 25). Compare with Posner (1970).

[99] White Paper on Modernisation of the Rules Implementing Articles 85 and 86 of the EC Treaty Commission Programme No 99/027 [1999] OJ C 132/1, para 57, emphasis added. Also Guidelines on the Application of Article 81(3) of the Treaty [2004] OJ C 101/97, House Of Lords Select Committee (2000: para 33, 60), Wesseling (1999: 424), Wesseling (2001: 372), and Proposal for A Council Regulation on the Implementation of the Rules on Competition Laid Down in Articles 81 and 82 of the Treaty and Amending Regulations (EEC) No 1017/68, (EEC) No 2988/74, (EEC) No 4056/86 and (EEC) No 3975/87 [2000] OJ C 365 E/284, 4. Though see Case T-193/03 *Lauren Piau v Commission* [2005] 5 CMLR 2, para. 100–104.

[100] Joined Cases T-374/94, T-375/94, T-384/94, T-388/94 *European Night Services v Commission* [1998] ECR II-3141, para 22–24, 107, 111, 113, 119, 132.

[101] Joined Cases T-374/94, T-375/94, T-384/94, T-388/94 *European Night Services v Commission* [1998] ECR II-3141, para 130.

[102] Joined Cases T-374/94, T-375/94, T-384/94, T-388/94 *European Night Services v Commission* [1998] ECR II-3141, para 136.

[103] Odudu (2002a).

[104] Case T-112/99 *Métropole Télévision (M6) v Commission* [2001] ECR II-2459, para 18–19, 49–66 and TPS [1999] OJ L90/6, para 102–109.

[105] Case T-112/99 *Métropole Télévision (M6) v Commission* [2001] ECR II-2459, para 68–69, 72.

the parties' claim.[106] Endorsing the Commission view, the CFI denied that balancing of productive and allocative efficiencies could take place under Article 81(1) EC as such an approach is inconsistent with the Treaty:

Article [81] of the Treaty expressly provides, in its third paragraph, for the possibility of exempting agreements that restrict competition where they satisfy a number of conditions... It is only in the precise framework of that provision that the pro and anti-competitive aspects of a restriction may be weighed... Article [81(3)] of the Treaty would lose much of its effectiveness if such an examination had to be carried out already under Article [81(1)] of the Treaty.[107]

Productive efficiency is a relevant consideration, and relevant only under Article 81(3) EC.[108]

B. *Assessing productive efficiency*

Having identified productive efficiency as the second relevant issue in an Article 81 EC enquiry a number of issues relevant to the assessment must be considered. First, whilst a practice may enable the colluding undertakings to improve their productive efficiency, the concern is with productive efficiency of the market rather than the particular undertakings.[109] This is confirmed by the Court in *Consten and Grundig*, holding that:

whether there is an improvement in the production or distribution of the goods in question... *cannot be identified with all the advantages which the parties to the agreement obtain from it in their production or distribution activities.* ... This subjective method, which makes the content of the concept of 'improvement' depend upon the special features of the contractual relationships in question, is not consistent with the aims of Article [81].[110]

Second, an assessment of productive efficiency may rely on information in the hands of the undertakings restricting competition, which may present the information selectively to distort the true picture.[111] This problem

[106] Case T-112/99 *Métropole Télévision (M6) v Commission* [2001] ECR II-2459, para 72.

[107] Case T-112/99 *Métropole Télévision (M6) v Commission* [2001] ECR II-2459, para 74. Also Robertson (2002: 98–100).

[108] Joined Cases T-374/94, T-375/94, T-384/94, T-388/94 *European Night Services v Commission* [1998] ECR II-3141, para 230 and European Night Services [1994] OJ L 259/20, para 59–61.

[109] Lipsey *et al.* (1993: 287), Williamson (1969b: 107), Williamson (1968a: 27), and Rose (1993: para 3-027).

[110] Cases 56 and 58–64 *Établissements Consten Sàrl and Grundig-Verkaufs-GmbH v Commission* [1966] ECR 299, 348, emphasis added.

[111] Kattan (1996: 618), Williamson (1977: 703), and Camesasca (1999: 22).

is countered in two ways: First, the burden of proving that market power is coupled with a positive effect on productive efficiency is on the party seeking to benefit from the claim.[112] Second, to substantiate the claim of a positive affect on productive efficiency requires 'a response that doesn't merely speculate about the existence of efficiencies, but rather comes forward with real-world evidence—factual evidence, expert economic evidence, and preferably both—to support the claim'.[113] Claims must not only be plausible; they must be credible. Moreover, the credibility of a plausible productive efficiency claim is affected by how the market power is obtained. For example, when conduct infringes Article 81(1) EC because the parties intend to restrict competition, the claim that there will be productive efficiencies does not seem credible.[114] The lack of credibility goes some way to explaining why collusion with the object of restricting competition rarely benefits from Article 81(3) EC, even though there are no infringements that *a priori* are incapable of satisfying the Article 81(3) EC criteria.[115]

Finally, Article 81(3) EC is said to involve 'extremely indeterminate legal concepts, which require complex economic assessments'.[116] National courts and competition authorities have had little opportunity to apply Article 81(3) EC, and even the Court of Justice has only considered Article 81(3) EC in the context of judicial review.[117] This has raised the question of whether national courts will be up to the task.[118] The concern is set aside because courts in various jurisdictions make similar assessments and

[112] Article 2 of Council Regulation (EC) No 1/2003 [2003] OJ L 1/1, Cases 1035/1/1/04 and 1041/2/1/04 *The Racecourse Association and Others v Office of Fair Trading* [2005] CAT 29, para 132–135, Deringer (1968: para 406), and Williamson (1968a: 24). The same position is taken in Canada Docket A-533-00 *Canada (Commissioner of Competition) v Superior Propane Inc. and ICG Propane Inc.* [2001] Carswell National 702, para 17, 143–154 and under the ECMR: Levy (2003: para 15.02(3)). Imposing penalties for providing false or deliberately misleading information deters spurious claims.

[113] Joel Klein, 'A Stepwise Approach to Antitrust Review of Horizontal Agreements', available at http://www.usdoj.gov/atr/public/speeches/jikaba.htm. Also Guidelines on Vertical Restraints [2000] OJ C291/1, para 136, Guidelines on the Application of Article 81(3) of the Treaty [2004] OJ C 101/97, para 51, 56–58, Guidelines on the Applicability of Article 81 of the EC Treaty to Horizontal Cooperation Agreements [2001] OJ C 3/2, para 32, ICN Merger Analytical Framework Subgroup (2004: para 87–90), and Levy (2003: para 15.02(4)(c)).

[114] Guidelines on Vertical Restraints [2000] OJ C 291/1, para 46–47 and Korah and O'Sullivan (2002: para 2.8.5, 4.2.1, 4.2.5–4.2.6, 4.6.4).

[115] Case T-17/1993 *Matra Hachette SA v Commission* [1994] ECR II-595, para 85, Boscheck (2000: 40), and Van den Bergh and Camesasca (2001: 248).

[116] Monopolkommission (2000: 19–20).

[117] Joined Cases T-374/94, T-375/94, T-384/94, T-388/94 *European Night Services v Commission* [1998] ECR II-3141, para 53, Organisation for Economic Co-operation and Development (1997: 55).

[118] Dibadj (2004: 782).

national courts already apply provisions involving similar analysis.[119] However, it remains necessary to specify how it is that the courts can, will, or ought to make the assessment of plausible claims that the ability to profitably contrive scarcity of output is a by-product of, or necessary element to, obtaining productive efficiencies. A strong argument can be made that productive efficiencies should be established in the same way that allocative inefficiencies are established in a particular case. For example, if the Article 81(1) EC infringement is established by measurement of allocative inefficiency, then Article 81(3) EC should require measurement of productive efficiencies.[120] The ease with which productive efficiencies are subject to measurement, prediction, and presumption is considered below.

1. Measurement

Bork argues that the quantification of productive efficiencies is too complex to be contemplated, arguing that whatever techniques are developed 'are spurious. They cannot measure the factors relevant to consumer welfare, so that after the economic extravaganza was completed we should know no more than before it began.'[121] However, productive efficiencies can be substantiated by measurement *ex post*. In *Maison Jallatte*, the Commission recognized that exclusive territories gave the distributor an incentive to promote the product. The productive efficiency benefits were quantified *ex post*, removing all ambiguity as to whether they would accrue. The promotional effort increased demand which resulted in 'a better spreading of the manufacturer's fixed costs, which is *confirmed by the sales figures known in this case*' and the savings from exploiting economies of scale 'have allowed both the manufacturer and the exclusive concession-holders to improve their cost prices and to transfer to the consumers a part of this advantage'.[122] In the final analysis, the process of quantifying productive efficiencies is really a qualitative one, 'not readily subjected to quantification or firm guidelines, not unlike judgments made by judges and fact-finders in many other areas of law'.[123] To do anything else is described as 'inherently

[119] Monti (2000: 3–4), Organisation for Economic Co-operation and Development (1997: 24), Organisation for Economic Co-operation and Development (1997: 27–29), Gyselen (2002: 187–188), Monti (2000: 3), Venit (2003: 560–561), Ehlermann and Atanasiu (2002: 75), and Fox (2001: 143).

[120] Compare with Patterson (2000: 445–447).

[121] Bork (1993: 124, also 123, 126–127), Williamson (1969b: 106), Van den Bergh and Camesasca (2001: 208), and American Bar Association (Antitrust Section) and Hartley (chairman) (1999: 134–135).

[122] *Maison Jallatte* [1966] JO 37/66 [1966] CMLR D1, D4–5, emphasis added.

[123] American Bar Association (Antitrust Section) and Hartley (chairman) (1999: 161, also 125).

arbitrary, because the factors cannot be measured and compared in any objective or mathematically reliable way'.[124]

2. Prediction

When collusion is unimplemented it is difficult to assess whether and to what extent benefits will accrue.[125] Merger control relies on the ability to make predictions about the productive efficiency consequences of integration and similar techniques can be applied to make predictions as to the productive efficiency consequences of collusion.[126] The likelihood of productive efficiency gains can be extraordinarily difficult to predict, particularly as so much depends on 'the intangible capital embodied in the work force'.[127] Katten thus asks '[h]ow can efficiencies have a central role in the analysis if specious, but plausible, efficiency claims are easy to assert and difficult to disprove while many valid claims are exceedingly difficult to prove, if they are at all susceptible to proof?'[128] Documents produced to discuss research and development activities and corporate strategies may provide a source for information on productive efficiencies.[129] This information can be combined with company financial documents to facilitate an examination into changes in various cost categories and output.[130] The probative value of such documents depends on the reason for which they were prepared and whether action was taken because of recommendations included in the documents.[131]

3. Presumption

Since, in determining productive efficiencies, 'some factors can be viewed as particularly significant', it is possible to develop a series of presumptions as to whether productive efficiencies are likely to exist.[132] This is the approach taken with the block exemption regulations, which establish a presumption of productive efficiency if certain conditions are met. The presumed productive efficiencies leading to the enactment of a block exemption are generally stated in the recitals; for example, in the fifth recital of Regulation 2790/1999 the aim of the exemption is considered to be to identify those situations in 'which it can be assumed with sufficient

[124] American Bar Association (Antitrust Section) and Hartley (chairman) (1999: 161).

[125] Ritter *et al.* (2000: 117).

[126] Camesasca (1999: 19), Kattan (1996: 619), and Scherer (2001: 19–23).

[127] Kendrick (1977: 69, also 3–4, 70–71) and Bork (1993: 127).

[128] Kattan, (1996: 614). [129] Kaplan (1998: 114–115). [130] Kaplan (1998: 110).

[131] Kaplan (1998: 114–115).

[132] American Bar Association (Antitrust Section) and Hartley (chairman) (1999: 161).

certainty that they satisfy the conditions of Article 81(3)'.[133] These presumptions are clearly developed from experience of similar types of collusion.[134] When the conditions of Article 81(3) EC cannot be satisfied, even when the terms of Regulation 2790/1999 are met, the Commission may withdraw the benefit of the block exemption, showing the presumption to be rebuttable.[135]

C. *The trade-off with allocative efficiency*

The relevance of productive efficiency in Article 81 EC is explained with the help of Figure 6.1.[136] As argued in Chapter 5, Article 81(1) EC is infringed when collusion is aimed at or causes allocative inefficiency. A contrived scarcity of output (Q_c to Q_a) results in an increased price (P_c to P_a) and causes those willing to pay the cost of production (P_c, which includes normal profit), but less than the sale price (P_a) to be denied consumption. This allocative inefficiency is marked as area A. It has been argued in Sections II and III above that productive efficiency is also a relevant competition law concern and in Sections IV.A and IV.B above that Article 81(3) EC instructs the decision making authority to carry out an

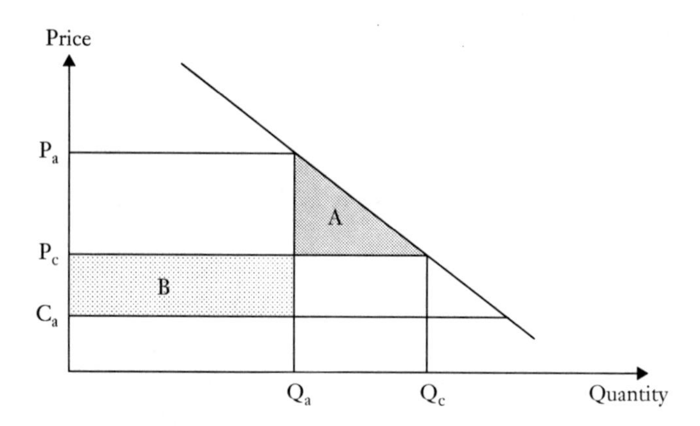

Figure 6.1 The efficiency trade-off.

[133] Commission Regulation (EC) No 2790/1999 [1999] OJ L 336/21, fifth recital.
[134] Goyder (1998: 565–566).
[135] Commission Regulation (EC) No 2790/1999 [1999] OJ L 336/21, recital 13 and Articles 6–8, Guidelines on Vertical Restraints [2000] OJ C 291/1, para 71–87, Whish (2000b: 918–921), Subiotto and Amato (2000: 15), and Subiotto and Amato (2001: 183–185).
[136] Compare with Williamson (1968a).

enquiry into the productive efficiency consequences of collusion with the aim or consequence of contrived scarcity of output. Productive efficiency gains are represented as a fall in the cost of production (synonymous with an increase in innovation) from P_c to C_a.

Determining the extent of any actual or expected productive efficiency gain is the first stage of an Article 81(3) EC analysis.[137] Any productive efficiencies shown to exist (area B) are weighed against allocative efficiency losses identified under Article 81(1) EC (area A).[138] Article 81(3) EC allows Article 81(1) EC to be declared inapplicable when there are countervailing productive efficiency gains.[139] In *Consten and Grundig*, the Court held that the productive efficiency gains should 'present noticeable objective advantages such as to compensate for the inconveniences resulting therefrom on the level of competition'.[140] This is generally taken to mean that the productive efficiency gain must outweigh the allocative efficiency loss.[141] This demarcation answers critics who consider it illogical to exempt collusion prohibited as anti-competitive under Article 81(1) EC because it promotes competition under Article 81(3) EC.[142]

A simple net gain is insufficient; Article 81(3) EC is also concerned with how the productive efficiency gains are distributed.[143] The decision to confer the benefit of Article 81(3) EC on the collusion of undertakings has consequences for the distribution of wealth across society and are illustrated in Figure 6.2. An Article 81(1) EC infringement occurs when contrived scarcity of output (Q_c to Q_a) and the consequent price increase (P_c to P_a) results in consumers willing and able to pay the cost of production being denied consumption (area A). The price increase also extracts more of what those who can still afford the goods and services are willing to pay (area C).[144] The benefit of Article 81(3) EC can be conferred if productive efficiency gains (area B) outweigh the allocative efficiency loss

[137] See Section IV.B.

[138] Compare with ICN Merger Analytical Framework Subgroup (2004: para 65–86).

[139] Manzini (2002). This does not suggest that competition authorities should act as management consultancies: Bork (1993: 123) and Neven *et al.* (1998: 20–21, 35).

[140] Cases 56 and 58–64 *Établissements Consten Sàrl and Grundig-Verkaufs-GmbH v Commission* [1966] ECR 299, 348.

[141] Guidelines on the Application of Article 81(3) of the Treaty [2004] OJ C 101/97, para 43, 85, Whish (2001: 125), and Van den Bergh and Camesasca (2001: 252). On the difficulties of achieving this balance see Williamson (1977: 702, 728), Camesasca (1999: 16–17), and American Bar Association (Antitrust Section) and Hartley (chairman) (1999: 124–126).

[142] Fox (2001: 142 note 6), Wesseling (2001: 359), and Monti (2002: 1060–1061).

[143] Collins (1984: 502). [144] Collins (1984: 502–503, 508–515).

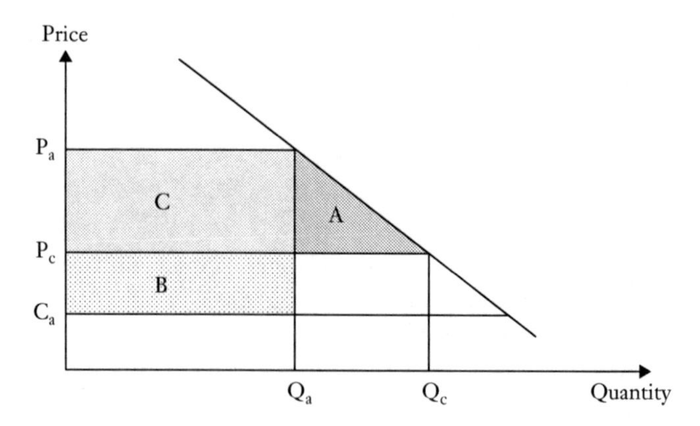

Figure 6.2 Distributional consequences of applying Article 81(3) EC.

(area A).[145] If area B is greater than area A, society on the whole is better off.[146] However, if exemption is not conferred, consumers are entitled to recover the overcharge (area C).[147] The application of Article 81(3) EC legitimizes the wealth transfer and prevents recovery. The question is whether, how, and why the distributional consequences illustrated by area C are relevant in the decision as to whether to confer the benefit of Article 81(3) EC on collusive conduct.[148] As stated by Collins, 'a political question arises of whether certain already allocated pieces can be recalled in order to make the pie larger'.[149] In Section IV.C.1 it will be argued that the distributional consequences of applying Article 81(3) EC are relevant because it is consumers that must pay more or forgo consumption in order for the productive efficiency gains to be produced. In Section IV.C.2 issues that arise in determining distribution are considered.

1. *The importance of distribution*

Pareto efficiency is achieved when nobody can be made better off without making somebody worse off.[150] Pareto efficiency is attractive as a goal because '*everyone* can be made better off if society is organized in an efficient

[145] See references cited at notes 140–141 above.
[146] Guidelines on the Application of Article 81(3) of the Treaty [2004] OJ C 101/97, para 85.
[147] Reich (2005) and Komninos (2002).
[148] Compare with Lande (1989), Feldman (1971: 516–517), and Markovits (1975b: 986).
[149] Collins (1984: 508). Also Ahdar (2002: 342, 352). Compare with Harrison (2004: 861).
[150] Posner (1981a: 88) and Hausman and McPherson (1996: 84).

manner'.[151] Since no one is made worse off, no one will object to an efficiency increase. The problem is that 'true Pareto improvements are rare: Economic changes usually involve winners and losers.'[152] One view is that competition law should not concern itself with the identity of the winners and losers, but simply ensure that there are more gains than losses.[153] This approach is encompassed in Harberger's third postulate:

> when evaluating the net benefits or costs of a given action (project, program, or policy), the costs and benefits accruing to each member of the relevant group (*e.g.*, a nation) should normally be added without regard to the individual(s) to whom they accrue.[154]

The rationale for ignoring the identity of those that gain and those that lose is that:

> 'any program or project that is subjected to applied-welfare-economic analysis is likely to have characteristics upon which *the economist as such is not professionally qualified to pronounce*... These elements—which surely include income-distributional... aspects of any project or program... may be exceedingly important, perhaps the dominant factors governing any policy decision, but they are *not part of the package of expertise that distinguishes the professional economist from the rest of humanity*. And that is why we cannot be expected to reach a professional consensus concerning them.'[155]

It is clear that distributional consequences matter, but are either too complex for economics to consider, or involve more issues than economists as economists are competent to consider. However, at times Harberger's third postulate metamorphoses from a claim that distribution is too complex for economics to a claim that distribution does not matter.[156] Distribution is de-emphasized by replacing the Pareto condition with the notion of potential-Pareto improvements (also known as Kaldor/Hicks improvements).[157] Rather than require nobody to be worse off and at least one

[151] Polinsky (1989: 7, emphasis added). Distributional concerns limit the normative appeal of efficiency: Areeda and Kaplow (1997: 8), Bator (1957: 28–29, 34–36, 39), Knight (1935: ch IX), Posner (1990: 357), Zweig (1980: 21–34), Böhm (1989: 53–54), Homan (1956: 25–27), and Carchedi (2001: 44–59).

[152] Hausman and McPherson (1996: 88).

[153] Kaplow and Shavell (2002), Polinsky (1989: 7), and Docket A-533-00 *Canada (Commissioner of Competition) v Superior Propane Inc. And ICG Propane Inc.* [2001] Carswell National 702, para 22–37.

[154] Harberger (1971: 785).

[155] Harberger (1971: 785–786, emphasis added, also 787).

[156] Amato (1997: 2–3), Bork (1993: 20–21, 90–106, 111), Hovenkamp (1982), Korah (2000: 11–12, 35), Peritz (1989), Posner (1976: 18–22), Van den Bergh and Camesasca (2001: 253–254), Williamson (1968a: 27–29), Williamson (1969b: 108–09), Williamson (1977: 711). *Contra* Ahdar (2002: 345–347) and Hughes (1994: 280–281).

[157] Hausman and McPherson (1996: 93–99).

person better off, the potential-Pareto criterion requires those that gain to be *able* to compensate those that lose and still be better off themselves.[158] If re-distribution *can be* made from the efficiency increase then no conflict between efficiency and distribution arises.[159] Whether redistribution is actually made is a political question, but if desirable many claim that it should be pursued through the tax system, which is a more accountable and effective way to achieve distributional aims.[160]

One reason to re-emphasize distribution is a paradox of its irrelevance. If distribution were not considered, those with a greater ability to profitably contrive scarcity of output are those more likely to gain exemption:

> This is because . . . purchasers of goods for which demand is inelastic are relatively insensitive to price [and so] fewer will purchase substitute goods despite increases in price. Therefore, a significant price increase will result in a smaller deadweight loss in a product where demand is inelastic than where it is elastic. . . . the more inelastic the demand for the goods produced . . . the smaller will be the efficiencies required . . . in order to offset its anti-competitive effects.[161]

A second reason to re-emphasize distribution is that the tax system is not always the most effective way to attain redistribution, particularly when the winners and losers are in different jurisdictions.[162] Finally, an increase in welfare that requires some to make sacrifices but not to benefit from the sacrifice, or worse, for the benefits to accrue to those that have not made sacrifices at the expense of those that have, seems to lack normative appeal. The appeal of Pareto efficiency is that nobody is made worse off by the change. The appeal of potential-Pareto efficiency is that compensation *can be* paid.[163] If compensation *is not* paid, efficiency seems to lack normative appeal.[164]

In the end, whether the losers 'should in fact be given compensation or not, is a political question'.[165] In Article 81 EC that question is answered affirmatively by the wording of Article 81(3) EC, which instructs the decision-making authority to be guided by a second positive condition that

[158] Hausman and McPherson (1996: 96–97), Posner (1981a: 91), and Jacobs (1995: 230).

[159] Polinsky (1989: 7–10).

[160] Kaldor (1939: 550), Bork (1965: 839), Kaplow and Shavell (2002), and Polinsky (1989: 119–130). *Contra* Collins (1984: 512–515) and Markovits (1975b: 988–989). On distortions caused by redistribution: Harberger (1978), Layard (1980), Squire (1980), and Harberger (1980).

[161] Docket A-533-00 *Canada (Commissioner of Competition) v Superior Propane Inc. And ICG Propane Inc.* [2001] Carswell National 702, para 104–105.

[162] Collins (1984: 512–515).

[163] Otherwise the criterion is defended as an analytical convenience: Posner (1981a: 92) and Rose-Ackerman (1992: 18–19).

[164] Markovits (1975b: 955–960), Duhamel and Townley (2003: 8–9), Ahdar (2002: 353) and Clark (1961: 77–80). For an attempt to justify the absence of a need to compensate: Posner (1981a: 94–103).

[165] Kaldor (1939: 550).

consumers receive a fair share of the benefit identified under the first positive condition—productive efficiency.[166] The distributional consequences of the infringement must be considered in deciding whether or not the benefit of Article 81(3) EC can be conferred. The first positive condition of Article 81(3) EC is satisfied if the increased price in the current period will allow the producers to lower their costs or innovate in future periods. Consumers face increased cost or are denied consumption if Article 81(3) EC is applied. Consumers are being asked to invest in future cost reductions and innovation (area B) by paying more for current consumption (area C). Allocative inefficiency and forgone consumption (area A) is an undesirable but necessary cost. Since it is consumers that invest in the creation of productive inefficiencies it is consumers that must benefit from this investment. This is ensured by the second positive condition preventing the application of Article 81(3) EC unless *consumers* benefit from the long-term gains.

2. *Determining the consumer's fair share*

Consumers are entitled to a fair share of the productive efficiency gains because it is consumers that must pay more or forgo consumption in order for the productive efficiency gains to be produced. In the past, the Court of Justice and Commission have proceeded on the assumption that consumers gain as long as the market remains sufficiently competitive.[167] However, as Article 81(3) EC is decentralized and productive efficiency claims are more clearly articulated, the importance of a more rigorous approach to the fair share criterion increases. Van den Bergh and Camesasca thus demand 'clear guidance on how much weight should be given to the requirement that efficiency gains should be passed on and under which timescale'.[168]

[166] On the origin of the idea that consumers should get a fair share of the benefits of competition: Deringer (1968: para 430, n 50), and Gerber (2001: 16–68). Compare with the treatment of consumer benefit under the ECMR efficiency defence: Levy (2003: para 15.02(4)(a)). The question of whether distribution is a reason for intervention is separate from the question of whether distributional issues arise. It seems that those arguing that distribution is not a concern argue that it is not a reason for intervention: Castañeda (1998: 46) Wolf (1998: 131–132). However, there is recognition that distributional consequences arise: Castañeda (1998: 49), Fels and Edwards (1998: 58, 63), and Schaub (1998: 120–123).

[167] Guidelines on the Applicability of Article 81 of the EC Treaty to Horizontal Cooperation Agreements [2001] OJ C3/2, para 34, Deringer (1968: para 438), Gyselen (2002: 184–185), Faull and Nikpay (1999: para 2.154–2.156), Buttigieg (2005: 706–717), Van den Bergh and Camesasca (2001: 208), Forrester QC (1998: 367–370), and Jones and Sufrin (2001: 196).

[168] Van den Bergh and Camesasca (2001: 320, also 3, 252), European Commission (2002), and Monti (2002).

One issue is identifying which consumers must benefit.[169] Article 81(3) EC requires that the consumers of the goods or services falling within Article 81(1) EC as opposed to the public in general should receive the benefit.[170] Further, it seems that the Treaty protects only Community consumers as opposed to consumers in general.[171] A condition for the infringement of Article 81(1) EC is that it affects trade between Member States, so it seems logical that the Article 81(3) EC benefits be felt where the infringement has effects. However, Community consumers still fall into two groups. There are first those paying the higher price (area C) and second those that could not afford to consume in the present period because of the increased price (area A). Although it is the second category of consumer that is acknowledged in Article 81(1) EC, Elzinga highlights the traditional failure of competition law to fully recognize and protect the position of the second category of consumer, writing that:

> [t]here are individuals, clearly singled out by economic theory as damaged in some amount, to whom the courts have not granted standing [to bring a private action]. An obvious example is the person who, because of a monopolistic price, does not buy the product but who would have purchased some quantity at a competitive price.[172]

Whether both the overcharged and those denied consumption count as consumers is relevant to a second issue—the form the benefit should take. Those who have been unable to consume the goods in the current period can be compensated with a lower price in the next period.[173] How much lower the future price should be can be calculated by applying a rate of interest to the value of forgone consumption for the period over which the benefit takes to accrue.[174] Those who have consumed in the current period,

[169] Guidelines on the Application of Article 81(3) of the Treaty [2004] OJ C 101/97, para 84.

[170] Guidelines on the Application of Article 81(3) of the Treaty [2004] OJ C 101/97, para 43, 86, Deringer (1968: para 424, 430–431), and Gyselen (2002: 185). Compare with the Australian position, described in Duns (1994: 253–256).

[171] This is the position in Australia: Duns (1994: 253). This issue under US antitrust is determined by the Foreign Trade Antitrust Improvements Act of 1982 (FTAIA), 15 USC § 6a, though the position is left somewhat unclear by *F Hoffman-La Roche Ltd v Empagran SA*, 542 US 155 (2004), discussed in Broder (2005: para 2-023, 30-50-3051). The issue has yet to be determined in Canada, Docket A-533-00 *Canada (Commissioner of Competition) v Superior Propane Inc and ICG Propane Inc.* [2001] Carswell National 702, para 168.

[172] Elzinga (1977: 1208–1209).

[173] Guidelines on the Application of Article 81(3) of the Treaty [2004] OJ C 101/97, para 95–101.

[174] Werden (1996) argues that the efficiency gains from a particular course of conduct can be measured and the effect of these gains on price predicted. Compare Guidelines on the Application of Article 81(3) of the Treaty [2004] OJ C 101/97, para 88.

but at a higher than competitive price, can be compensated by the availability of what it will be possible to produce in the future.[175] Whether this is a benefit to this class of consumers as a whole is ambiguous: infra-marginal consumers would prefer a lower price to the promotional effort, service, or innovation, and so are forced to pay for the provision of something they do not value.[176] Marginal consumers value promotional effort, service, or innovation. Whether or not the innovation is welfare-enhancing depends on the proportion of marginal to infra-marginal consumers.[177]

A third issue is of the timeframe in which the benefits must be conferred on consumers.[178] The realization of productive efficiency gains and the transmission of benefits to consumer must be reasonably proximate in time to the allocative efficiency sacrificed; the greater the time lag, the greater must be the compensatory benefits.[179]

D. *The negative conditions and their function*

Article 81(3) EC contains two positive conditions: first, there must be productive efficiency gains that outweigh the allocative efficiency loss.[180] Second, consumers must obtain enough of these productive efficiency gains to compensate them for their allocative efficiency loss.[181] However, there are difficulties in determining (a) whether productive efficiency gains exist, (b) whether the productive efficiency gains outweigh the allocative efficiency sacrificed to create them, and (c) whether consumers receive a fair share of the productive efficiency gains. It is because of these difficulties that Article 81(3) EC cannot be applied unless two negative conditions are also satisfied.[182] The first negative condition is that the restrictions causing allocative inefficiency must not be dispensable; the second negative condition

[175] Feldman (1971: 510) and Guidelines on the Application of Article 81(3) of the Treaty [2004] OJ C 101/97, para 86, 89, 102–104.

[176] Tirole (1988: 182), Carlton and Perloff (2000: 408–409), and Van den Bergh and Camesasca (2001: 219–220).

[177] Comanor (1968: 1429–1430), Comanor (1985: 990–999), Williamson (1979b: 966), and Neven *et al.* (1998: 21–24). This objection is considered in Kelly (1988: 371–374) and Mastromanolis (1995: 592–594).

[178] Deringer (1968: para 428) and Williamson (1969b: 106).

[179] Guidelines on the Application of Article 81(3) of the Treaty [2004] OJ C 101/97, para 87–88, Ahdar (2002: 351), Williamson (1968a: 25–26), Faull and Nikpay (1999: para 1.118–1.121), Vickers (1995: 7), and Levy (2003: para 15.02(4)(a)).

[180] See Section IV.C above.

[181] See Section IV.C.1. above.

[182] The cumulative nature of the conditions is confirmed in Guidelines on the Application of Article 81(3) of the Treaty [2004] OJ C 101/97, para 38, 42.

is that competition must not be eliminated. These conditions are explained in turn below, and it is argued that they minimize the consequences of error in assessing the two positive conditions.

1. No indispensable restrictions

The first negative condition in Article 81(3) EC requires those restricting competition within the meaning of Article 81(1) EC to refrain from 'impos[ing] on the undertakings concerned restrictions which are not indispensable to the attainment of these objectives'. The indispensability requirement is an aspect of the proportionality principle.[183] By ensuring that the allocative efficiency loss is minimized before the benefit of Article 81(3) EC is conferred, the cost of errors in determining (a) whether productive efficiency gains exist and (b) whether the productive efficiency gains outweigh the allocative efficiency sacrificed to create them, is minimized.

It has been argued that whilst not indispensable to achieving productive efficiency, restrictions within the meaning of Article 81(1) EC may be indispensable to achieving some other valuable goal, and that this can be considered.[184] However, it is clear that the restrictions within the meaning of Article 81(1) EC must be indispensable to achieving the benefits identified by the two positive conditions in Article 81(3) EC—productive efficiency, of which consumers receive a fair share.[185] In order for a restriction to be indispensable, there must be (i) a strong connection between the allocative efficiency loss and the productive efficiency gain and (ii) no obvious alternative methods available that are substantially less restrictive.[186] The condition is satisfied if the productive efficiency gain would not arise or would take significantly longer to arise in the absence of the allocative efficiency sacrifice.[187]

In assessing the indispensability of a restriction one question is whether it is for undertakings claiming the benefit of Article 81(3) EC to prove that

[183] Gyselen (2002: 186–187) and Whish (2001: 130).

[184] Bouterse (1994: 25) and Case 75/84 *Metro SB-Großmärkte GmbH & Co KG v Commission* [1986] ECR 3021, para 65.

[185] Deringer (1968: para 443), Guidelines on the Application of Article 81(3) of the Treaty [2004] OJ C 101/97, para 42, 73–82, and Faull and Nikpay (1999: para 2.165).

[186] Guidelines on the Application of Article 81(3) of the Treaty [2004] OJ C 101/97, 74–76 and Faull and Nikpay (1999: para 2.168). Compare with American Bar Association (Antitrust Section) and Hartley (chairman) (1999: 165–166), Hylton (2003: 121–124), and Roth QC (2001: para 3-055-3-058).

[187] Guidelines on the Application of Article 81(3) of the Treaty [2004] OJ C 101/97, para 74–75, 79, 89 and Faull and Nikpay (1999: para 2.166).

the restriction is indispensable. On its face Article 2 of Council Regulation 1/2003 places the burden of proving all the Article 81(3) EC elements on undertakings claiming the benefit of the provision.[188] However, it seems too great a burden to show the validity of a negative condition.[189] Moreover, the past practice has been for the Commission to show that certain restrictions are dispensable.[190] It may well be that there is a presumption that the condition is satisfied, rebuttable by the party that established the Article 81(1) EC infringement.[191]

2. No possibility of eliminating competition

The second negative condition in Article 81(3) EC ensures that Article 81(1) EC is enforced unless collusion does not 'afford such undertakings the possibility of eliminating competition in respect of a substantial part of the products in question'. The cost of error in determining whether consumers receive a fair share is minimized by this condition as it is assumed that consumers gain from productive efficiencies as long as the market remains sufficiently competitive.[192] This is illustrated by the Commission concern to examine whether consumers have in fact benefited in examining whether this condition is satisfied in relation to implemented collusive practices.[193]

The second negative condition in Article 81(3) EC speaks of *eliminating* competition, whilst Article 81(1) EC speaks of *restricting* competition.[194] In the abstract there is no way to determine when restricting competition becomes eliminating competition, so a margin of appreciation must lie with the decision maker.[195] The Commission has expressed the view that Article

[188] Council Regulation (EC) No 1/2003 [2003] OJ L 1/1 and Guidelines on the Application of Article 81(3) of the Treaty [2004] OJ C 101/97, para 41.

[189] American Bar Association (Antitrust Section) and Hartley (chairman) (1999: 123, 133–134).

[190] Joined Cases T-374/94, T-375/94, T-384/94, T-388/94 *European Night Services v Commission* [1998] ECR II-3141, para 210–215 and Whish (2001: 131).

[191] This is not consistent with Guidelines on the Application of Article 81(3) of the Treaty [2004] OJ C 101/97, para 78.

[192] Commission Regulation (EC) No 2790/1999 [1999] OJ L 336/21, recital 7–9, Guidelines on the Applicability of Article 81 of the EC Treaty to Horizontal Cooperation Agreements [2001] OJ C 3/2, para 34, Deringer (1968: para 438), Gyselen (2002: 184–185), Faull and Nikpay (1999: para 2.154–2.156), Buttigieg (2005: 706–717), Van den Bergh and Camesasca (2001: 208), Forrester QC (1998: 367–370), Whish (2000b: 908), and Jones and Sufrin (2001: 196).

[193] Guidelines on the Application of Article 81(3) of the Treaty [2004] OJ C 101/97, para 111.

[194] Jones and Sufrin (2001: 198) do not draw the distinction, so that Article 81(3) EC cannot be applied if the parties acquire market power. This renders Article 81(3) EC otiose since it is market power (sub-dominance) that establishes the Article 81(1) EC infringement.

[195] Bailey (2004: 1337–1339), Bouterse (1994: 24–26), and Whish and Sufrin (2000: 145).

81(3) EC will not be applied if collusion may eliminate competitors from the market place, writing that '[u]ltimately the protection of rivalry and the competitive process is given priority over potentially pro-competitive efficiency gains'.[196] This position must be seen in the context of the function the elimination condition serves—ensuring fair share to consumers—in order to prevent an interpretation that allows the benefit of Article 81(3) EC to be denied to protect inefficient competitors. The condition will not protect productively inefficient competitors if it is interpreted as simply requiring that collusion not result in the creation of additional barriers to entry or result in an unacceptable risk that additional barriers to entry will be erected.[197] To some extent this approach is taken.[198] The twelfth recital of Regulation 2790/1999 explains that the conditions of a block exemption 'normally ensure that the agreements to which the block exemption applies do not enable the participating undertakings to eliminate competition in respect of a substantial part of the products in question'.[199] One condition, laid down in Article 3 of Regulation 2790/1999, is that the colluding undertakings should not have a market share exceeding 30 percent: 'Above that level there is a risk that the last condition of Article [81(3)] is no longer fulfilled.'[200] The market share of agreement participants, the type of product in question, the number and strength of the remaining competitors, and the existence of potential competition have all been examined to ensure that competition is secure, despite the allocative inefficiency that a declaration of inapplicability will allow to persist.[201] Without creating insurmountable entry barriers or increasing the risk that such barriers will be created, it is difficult to see how an agreement creates the risk of eliminating competition.[202] However, as Faull & Nikpay note, 'Commission decisions have not always given this

[196] Guidelines on the Application of Article 81(3) of the Treaty [2004] OJ C 101/97, para 105.

[197] Gyselen (2002: 184) and Amato (1997: 109–113).

[198] Guidelines on the Application of Article 81(3) of the Treaty [2004] OJ C 101/97, para 114–115. Though compare with Guidelines on Vertical Restraints [2000] OJ C 291/1, para 135, Bishop and Ridyard (2002: 36–37), Peeperkorn (2002: 39–40), Case C-2/1991 *Criminal Proceedings against Wolf W Meng* [1993] ECR I-5751, 5768–5770, and Mortelmans (2001: 620, 643–644, n 146).

[199] Commission Regulation (EC) No 2790/1999 [1999] OJ L 336/21, recital 12.

[200] Communication from the Commission on the Application of the Community Competition Rules to Vertical Restraints (Follow-up to the Green Paper on Vertical Restraints) Com(98)544 Final [1998] OJ C 365/3, 21–22.

[201] Guidelines on the Application of Article 81(3) of the Treaty [2004] OJ C 101/97, para 108–110, Faull and Nikpay (1999: para 2.172), and Rose (1993: para 3–045).

[202] Compare with Faull and Nikpay (1999: para 3.33, 3.49–3.72).

condition as much consideration as some of the other conditions of Article 81(3).'[203] So it remains to be seen how the interpretation of this condition develops.

V. Conclusion

It has been argued that Article 81(3) EC involves a productive efficiency enquiry. Section II explains why collusion causing allocative inefficiency, thus infringing Article 81(1) EC, may also cause productive inefficiencies, compounding the problem of anti-competitive conduct. Section III explains why collusion infringing Article 81(1) EC might also enhance productive efficiency. One aspect of productive efficiency is technological progress, which describes new and better goods and services being made available or familiar goods and services being produced in new and better ways. Arguments that supra-normal profits, and the allocative inefficiency this entails, are a necessary incentive to achieve this aspect of productive efficiency are considered. Both productive and allocative efficiency are relevant to competition law as both affect living standards. Section IV shows that Article 81(3) EC involves an enquiry as to whether the loss in allocative inefficiency identified under Article 81(1) EC is offset by a gain in productive efficiency. That productive efficiency is the focus of the enquiry; how productive efficiency is quantified; and how quantified productive efficiencies are balanced against allocative inefficiencies identified under Article 81(1) EC are explained. The first positive condition in Article 81(3) EC requires productive efficiencies to be shown and to outweigh the allocative inefficiencies identified under Article 81(1) EC. However, importance is attached to the distribution of gains as between producers and consumers. The distribution of identified productive efficiency gains is and ought to be tilted in favour of consumers by the second positive condition in Article 81(3) EC, as it is consumers that forgo allocative efficiency to generate the productive efficiency gains. Finally, Article 81(3) EC contains two negative conditions. The first is a proportionality requirement; Article 81(3) EC can only be applied when the Article 81(1) EC infringement is minimized to the greatest possible extent. This minimises both the loss that must be suffered to obtain the productive efficiency gains and the cost of error should the promised gains not materialize. The second

[203] Faull and Nikpay (1999: para 2.170).

negative condition prevents Article 81(3) EC applying if competition will be eliminated as a result. This should not be read as protecting competitors from their inefficiency, but instead prevents conditions arising that hinder competition on the merits.

Hawk expressed the view that 'the bifurcation of what ideally should be a single anti-trust analysis into the double tests of Article [81(1)] and Article [81(3)]' results in 'near anarchy'.[204] Whish asks whether 'the bifurcation of Article [81 is] the "original sin" of European competition law?'[205] In arguing that Article 81(3) involves a productive efficiency enquiry, the coherence and underlying logic of the bifurcated analysis contained in Article 81 is revealed. Article 81(1) EC involves an allocative efficiency enquiry and a restriction of competition involves causing allocative inefficiency. Article 81(3) EC involves a productive efficiency enquiry; allocative inefficiency can be invested or forgone to incentivize the creation of productive efficiencies. This answers those that consider it illogical to exempt collusion prohibited as anti-competitive under Article 81(1) EC because it promotes competition under Article 81(3) EC.[206] Finally, it is right that consumers benefit from the productive efficiency gains, as it is they that have forgone consumption and paid supra-normal prices necessary to create them.

[204] Hawk (1988: 69). [205] Panel Discussion (1998: 462).

[206] Fox (2001: 142 n 6), Wesseling (2001: 359), and Monti (2002: 1060–1061).

7

Article 81 EC and non-efficiency goals

I. Introduction

Chapters 5 and 6 have argued that, as a substantive matter, Article 81 EC is concerned with efficiency. Article 81(1) EC is infringed by collusion with the aim or consequence of reduced output and increased price—allocative inefficiency. Article 81(3) EC allows Article 81(1) EC to be declared inapplicable when the collusion also results in productive efficiency, so that net efficiencies are considered. The purpose of this chapter is to show efficiency as the exclusive concern. Areeda and Kaplow note that '[t]he efficiency concept is at once both powerful and weak: powerful because it is arguably the minimum necessary condition of any ideal economic system's equilibrium, weak because it is only one of many values considered important by our society.'[1] Section II acknowledges that non-efficiency goals have, in the past, been considered in Article 81 EC. Section III argues that in the future non-efficiency goals should not be considered in an Article 81 EC enquiry: it was always inappropriate to use Article 81 EC to advance these goals; legal certainty and justiciability are enhanced by a unitary goal; and, a unitary goal maintains the intended relationship between Article 81 and other Treaty provisions. Section IV concludes that the extent to which non-efficiency goals trump efficiency is a constitutional question external to competition law that requires resolution at the constitutional level.

[1] Areeda and Kaplow (1997: 6).

II. The consideration of non-efficiency goals in Article 81 EC

It is clear that non-efficiency goals have been considered within Article 81 EC in the past. This results first from a teleological interpretation of Article 81 EC.[2] Article 2 EC sets out the objectives of the Community that are to be achieved by policies set out in Articles 3 EC and 4 EC. Article 3(1)(g) EC provides the basis for the competition rules, and is amplified in part by Article 81 EC.[3] Through a teleological interpretation of Article 81 EC, the Court endorsed pursuit of a range of Article 2 EC goals, particularly through the use of the Article 81(3) EC power to declare Article 81(1) EC inapplicable.[4] This view of the relationship between Article 81 EC and the remainder of the Treaty is expressed by Ellis:

> In order to assess the significance of article [81], it is essential to view it in relation to the Treaty as a whole and in the context in which it has been placed. Every provision fulfils a function in the general scheme of the Treaty, and it therefore seems opportune and useful to determine the function of article [81] in its context.[5]

The Commission has described the exemption procedure as involving 'balancing competition policy against other policies of the Community'.[6] Ehlermann, when Director General of Competition at the Commission, stated:

> It would probably be an exaggeration to assume that . . . non-economic considerations are to be totally excluded from the balancing test required by Article 81(3). Such an interpretation would hardly be compatible with the Treaty, the Court of Justice's case law, and the Commission's own practice.[7]

[2] Arnull *et al.* (2000: 539–541), Brown and Kennedy (2000: 334–337, 339–344), Albors-Llorens (1999: 381–383), Craig and de Búrca (2003: 98), and Koopmans (1986: 928–929).

[3] Wesseling (2000: 15–17).

[4] Case 26/76 *Metro SB-Großmärkte GmbH & Co KG v Commission* [1977] ECR 1875, para 21, Case 75/84 *Metro SB-Großmärkte GmbH & Co KG v Commission* [1986] ECR 3021, para 65, Bouterse (1994: 22, 42, 44, 47–51, 54–55), Cooke (2000: 61–63), Ehlermann (2000: 557), Ellis (1963: 263–265, 271–276), Fels and Edwards (1998: 54, 63–64), Gerber (2001: 417–420), Panel Discussion (1998: 469, 486–487, 480, 483–484, 489, 490–491), Peeters (1989: 521–523), Snyder (1990: 94–95), Wesseling (1999: 424–425), and Wesseling (2000: 94–96, 105–112).

[5] Ellis (1963: 248). Also Whish (1998: 495) and Wesseling (2000: 18).

[6] Green Paper on Vertical Restraints in EC Competition Policy Com (96) 721 Final [1997] 4 CMLR 519, para 191.

[7] Ehlermann (2000: 549). Also Ehlermann (1995: 479–481), Van den Bergh and Camesasca (2001: 159–165), and Prosser (2005: 24–28).

In addition to teleology, non-efficiency factors are considered in Article 81 EC because the Community agenda has evolved from the creation of an economic Community to the creation of a social or citizens' Community. The result of this 'horizontal expansion' is the addition of various policies that must be pursued simultaneously with existing policies, including competition law.[8] Articles 6 EC, 127(2) EC, 151(4) EC, 152(1) EC, 153(2) EC, 159 EC and 178 EC, termed 'policy-linking', 'policy-integration', or 'flanking' clauses, are provisions in the EC Treaty that are to be considered in the definition and implementation of all other Community policies.

A combination of the teleological approach and horizontal expansion has resulted in an orientation of Article 81 EC 'directed at promoting the various other objectives of the Community enshrined in Article 2 EC'.[9] In this regard, Joliet was mistaken in considering that Article 81(3) EC 'strictly codified the tests which are to be followed by the Courts or the enforcement agency in balancing good and bad restraints'.[10] A regularly cited example of balancing competing and conflicting objectives *within* Article 81(3) EC is the Commission's examination of agreements notified by the European Committee of Domestic Equipment Manufacturers (CECED), to which 90–95 per cent of European dishwasher manufacturers are party.[11] Dishwasher manufacturers agreed to stop producing or importing high electricity-consuming appliances, affecting 63 per cent of such goods

[8] Dougan (2000: 856–859) and Weiler (1999: 39–63).

[9] Wesseling (2000: 49). Also Case 75/84 *Metro SB-Großmärkte GmbH & Co KG v Commission* [1986] ECR 3021, para 65, Joined Cases T-528/1993, T-542/1993, T-543/1993 and T-546/1993 *Metropole Télévision SA v Commission (Eurovision)* [1996] ECR II-649, para 112, Case C-309/99 *Wouters v Algemene Raad Van de Nederlandse Orde Van Advocaten (Raad Van de Balies Van de Europese Gemeenschap, Intervening)* [2002] ECR I-1577, AG Opinion para 113, Guidelines on the Applicability of Article 81 of the EC Treaty to Horizontal Cooperation Agreements [2001] OJ C3/2, para 179–197, Banks (1997), Bengoetxea *et al.* (2001: 67–81), Bishop and Bishop (1996: 208), Bishop (2000: 43), Black (1997: 148), Bouterse (1994: 41, 45–46, 52–54), Cooke (2000: 67), Cosma and Whish (2003: 41–42), Gaay Fortman (1966: 151–158), Gerber (1994: 131–132), Hawk (1988: 56), Hawk (1995: 987–988), Hildebrand (1998: 12–19), Hornsby (1987), House of Lords Select Committee (2000: para 54, 58, 130–132, 149), Marsden (1997: 237), McNutt (2000: 50, 56–58), Monti (2002: 1069–1078), Schmid (2000: 167–171), Siragusa (1998: 665–666), Siragusa (2000: 1101), Wesseling (1999: 423–424), Wesseling (2000: 18, 33, 61, 33–48, 62–64, 83–88, 103–105), Whish (1997), Whish and Sufrin (2000: 147–149), and Whish (2001: 95).

[10] Joliet (1967: 6). Also Ritter *et al.* (2000: 116) and Deringer (1968: para 423).

[11] *CECED* [2000] OJ L187/47, para 8, 24. Similarly see Case C-67/96 *Albany International BV v Stichting Bedrijfspensioenfonds Textielindustrie* [1999] ECR I-5751, AG Opinion para 126, 193, Case 26/76 *Metro SB-Großmärkte GmbH & Co KG v Commission* [1977] ECR 1875, para 43, Case 42/84 *Remia BV v Commission* [1985] ECR 2545, para 42, *Synthetic Fibres* OJ 1984 L 207/17, para 37, *Ford/Volkswagen* OJ 1993 L 20/14, para 23, 36, and Boni and Manzini (2001: 244).

currently sold.[12] Further, they agreed to pursue joint energy efficiency targets, improve consumer information on energy-efficient use and to develop more environmentally friendly machines.[13] The stated aim of the agreements was to improve the energy efficiency of appliances marketed in the European Union. However, energy-efficient washing machines are more costly to produce and would result in a price increase of up to 14 per cent.[14] Collusion with the aim or consequence of reduced output and the corollary of increased price is an infringement of Article 81(1) EC.[15] The Commission thus considered the Article 81(3) EC conditions, finding that 'the new electric water heaters and dishwashers will be more efficient and enable consumers to reduce their energy bills. Moreover, *lower electricity consumption will indirectly help the Union achieve its environmental objectives*.'[16] There is a view that exemption was granted in part because '[r]educed electricity consumption indirectly leads to reduced pollution from electricity generation'.[17]

In addition to the pursuit of non-efficiency aims under Article 81(3) EC, there is also a claim that non-efficiency aims are relevant under Article 81(1) EC. In *Albany* Article 81(1) EC was said not to apply to collective bargaining between management and labour on wages and working conditions.[18] Article 2 EC and other Treaty provisions require 'the raising of the standard of living' and the application of Article 81(1) EC might prevent workers being paid higher wages.[19] In *Wouters*, the Nederlandse Orde van Advocaten, the Netherlands Bar Association (NBA), composed of all lawyers registered in the Netherlands, was established by statute and empowered to regulate the legal profession. Though lawyers could enter

[12] *CECED* [2000] OJ L 187/47, para 45.

[13] *CECED* [2000] OJ L 187/47, para 19–22.

[14] *CECED* [2000] OJ L 187/47, para 16–17, 34.

[15] *CECED* [2000] OJ L 187/47, para 33.

[16] Commission Press Release IP/01/1659, 26 November 2001, emphasis added.

[17] *CECED* [2000] OJ L187/47, para 48, also para 51. Also Commission Press Release IP/01/1659, 26 November 2001, Case 240/83 *Procureur de la Republique v ADBHU* [1985] ECR 531, para 11–15. The Commission rejects the idea that the non-economic environmental benefits were considered. Instead the economic savings from the reduced need to clean up the environment and lower running costs for consumers were considered: *CECED* [2000] OJ L187/47, para 11–12, 52, 56, Monti (2002: 1073–1077), Gyselen (2002: 185–186), and Wesseling (2001: 370–371).

[18] Case C-67/96 *Albany International BV v Stichting Bedrijfspensioenfonds Textielindustrie* [1999] ECR I-5751, para 46–60, AG Opinion para 120–194.

[19] Case C-67/96 *Albany International BV v Stichting Bedrijfspensioenfonds Textielindustrie* [1999] ECR I-5751, para 54–59, Boni and Manzini (2001: 244–245), Van den Bergh and Camesasca (2000: 502–508 and references cited at n 6–17 plus text accompanying), and Ichino (2001: 187).

into partnership with other lawyers registered in the Netherlands, other Member States, or third states, lawyers wishing to enter into partnership with members of a different profession required NBA authorization. Lawyers were authorized to enter into partnership with notaries, tax consultants, and patent agents, but were not authorized to enter into partnership with accountants.[20] Both Advocate General Léger and the Court found that preventing lawyers from entering partnerships with accountants prevented the creation of better services tailored to clients operating in a complex economic and legal environment and so had the effect of restricting competition.[21] Advocate General Léger considers Article 81(1) EC as concerned with 'a *purely competitive* balance-sheet of the effects of the agreement... The only legitimate goal which may be pursued in accordance with that provision is therefore exclusively competitive in nature.'[22] However, the Court ruled that in some circumstances collusion may pursue an *objective* that overrides the requirements of Article 81(1) EC and these *objectives* in the public interest are not covered by the competition rules at all, as long as the means are proportionate to the objective being attained.[23]

III. Rejecting non-efficiency considerations in Article 81 EC

In the summer of 1997, after noting that 'the Treaty itself is to be interpreted teleologically, and the text, from the Preamble, through Arts. 2 and 3, and on into the specific policies, is capable of supporting a variety of objectives', Whish asked whether it is:

desirable or possible that Arts. [81 and 82] should be read in a more detached way: should competition policy be isolated—from the (contaminating?) influence of

[20] The national legal framework is more fully set out by Deards (2002: 619).

[21] Case C-309/99 *Wouters v Algemene Raad Van de Nederlandse Orde Van Advocaten (Raad Van de Balies Van de Europese Gemeenschap, Intervening)* [2002] ECR I-1577, para 81–90, 95–96, AG Opinion para 42, 116–133. On the *object* of the collusion: para 43, 75–78, AG Opinion para 42, 94–95.

[22] Case C-309/99 *Wouters v Algemene Raad Van de Nederlandse Orde Van Advocaten (Raad Van de Balies Van de Europese Gemeenschap, Intervening)* [2002] ECR I-1577, AG Opinion para 104, emphasis added.

[23] Case C-309/99 *Wouters v Algemene Raad Van de Nederlandse Orde Van Advocaten (Raad Van de Balies Van de Europese Gemeenschap, Intervening)* [2002] ECR I-1577, para 97, 109–110, Andrews (2002), Deards (2002: 622), Denman (2002), Gyselen (2000: 435), and Vossestein (2002: 856–859).

other provisions of the Treaty? Is this legally possible? Why should competition provisions be singled out for this separate analysis and application?[24]

Amato advances the idea that Article 81 EC should be given autonomous analysis, writing that 'convergent indications are coming for a liberation of antitrust from the multiple purposes it has served in the past, enabling it to be . . . antitrust law pure and simple'.[25] There are a number of reasons why Article 81 EC should be liberated. It is argued first that the consideration of other Treaty goals within Article 81 EC is illegitimate. It is then argued that pursuit of a unitary goal renders Article 81 EC more effective.

A. *The illegitimacy of considering additional goals*

There are those that deny the legitimacy of balancing other Treaty goals within Article 81, since it requires 'reading into Art. [81(3)] something that does not appear in its wording, ie that unemployment, environment, or other such concerns can trump competition policy'.[26] The legitimacy of balancing efficiency against non-efficiency goals *within* Article 81 EC can be challenged on three grounds. First, the approach considers issues already considered in determining whether Article 81 EC was the appropriate rule to apply to the activity in question. Second, balancing efficiency against non-efficiency goals *within* Article 81 EC creates horizontal direct effect for provisions incapable of direct effect. Finally, specific Treaty provisions determining when efficiency is to be forsaken in favour of other objectives would be otiose if competing objectives were to be balanced within Article 81 EC. These challenges to the legitimacy of considering non-efficiency goals within Article 81 EC are made in turn.

1. *Reconsideration of jurisdictional questions*

When Treaty provisions apply to Member States, the Court is clear that it is possible to balance a range of objectives within the particular provision

[24] Whish (1998: 495–496).

[25] Amato (1997: 116). Compare with Case C-2/1991 *Criminal Proceedings against Wolf W Meng* [1993] ECR I-5751, 5770 and Jenny (1994: 195–197).

[26] Panel Discussion (1998: 478, also 485), Case C-67/96 *Albany International BV v Stichting Bedrijfspensioenfonds Textielindustrie* [1999] ECR I-5751, AG Opinion para 175, Case C-2/1991 *Criminal Proceedings against Wolf W Meng* [1993] ECR I-5751, 5768–5770, Joined cases T-528/1993, T-542/1993, T-543/1993 and T-546/1993 *Metropole Télévision SA v Commission (Eurovision)* [1996] ECR II-649, para 104–105, para 116–117, Mortelmans (2001: 620, 643–644, and note 146), Gyselen (2002: 185–186), and Monopolkommission (2000: 37 n 26).

being applied.[27] Importing this approach into Article 81 EC is supported so that the outcome of the case is not determined by the status of the actor.[28] Convergence of approach is required because privatization and contracting out have transferred tasks previously performed by the state to undertakings operating in the market place.[29] Since undertakings now perform the same functions as previously performed by the state, they should benefit from the same Treaty derogations as the state when performing those functions.[30]

It has been argued (Chapter 3) that Article 81 EC imposes obligations on the citizen (undertaking) rather than the state.[31] The possibility of balancing a range of Treaty goals *within* a single Treaty provision is specifically and exclusively designed for *Member States*, treating an infringement of the Treaty by the state more flexibly than an infringement of the Treaty by the citizen because state action has a democratic legitimacy and political accountability that private action lacks.[32] The problem of undertakings performing state activity is recognized by taking a *functional* approach to determine the addressees of Article 81 EC.[33] Chapter 3: Section IV.A.3 noted that the functional interpretation of undertaking allocated activities to be assessed under private law or public law norms—to competition or free-movement. If the activities are allocated correctly at the outset, there is no need to import public law flexibility into private law. Instead, activities requiring the flexibility of public law are allocated to the public law sphere.[34] Difficulties with the application of the functional approach have

[27] Case 8/74 *Procureur du Roi v Benoît and Gustave Dassonville* [1974] ECR 837, para 6, Case 120/1978 *Rewe-Zentral AG v Bundesmonopolverwaltung Für Branntwein* [1979] ECR 649, para 8, Case 140/1979 *Chemial Farmaceutici SpA v DAF SpA* [1981] ECR 1, para 13, 16, Joined Cases C-267/1991 and C-268/1991 *Criminal Proceedings against Bernard Keck and Daniel Mithouard* [1993] ECR 6097, paras 13–15, Craig and de Búrca (2003: 617–626, 659), Hatzopoulos (2000: 76–80), Weatherill and Beaumont (1999: 472–474, 521–524, 578–579), Schmid (2000: 166–167), Mortelmans (2001: 642), and Cruz (2002: 119, 121–125).

[28] Monti (2002: 1088–1089).

[29] Cruz (2002: 3–4, 91–104), Estrin (1994), Monti (2002: 1081, 1086–1090), Mortelmans (2001: 637, 640), O'Loughlin (2003), Prosser (2005: 20–24, 44–47), Wright (1994), and Wright and Perrotti (2000).

[30] Mortelmans (2001: 615–616, 638–642) and Hatzopoulos (2000: 65–74).

[31] Whish (2003: 211), van den Bogaert (2002: 123), and Mortelmans (2001: 622–623, 635–636).

[32] Bamforth (1997: 137, 140–151), Cruz (2002: 88–89, 111, 132–134, 149, 155–161), Loughlin (2003: 1, 5, 7–12), Maduro (1997: 55–56), Oliver (1999: 12–13, 19–22, 80–88), Snell (2003: 324, 337), and Stiglitz (2000: 13–14). Dougan (2000: 866–884), gives an account of flexibility predicated on the assumption that the flexibility is being accorded to the state.

[33] Mortelmans (2001: 624, 634–635), Woolf (1995: 63–64), and Cruz (2002: 87–88, 121–125).

[34] Case C-309/99 *Wouters v Algemene Raad Van de Nederlandse Orde Van Advocaten (Raad Van de Balies Van de Europese Gemeenschap, Intervening)* [2002] ECR I-1577, para 56, 70, AG Opinion

at times led the Court to conflate the jurisdictional question with questions of substance and justification.[35] Classification as public or private activity is then of less importance, since the correct norm is applied regardless of how the activity is initially classified. This approach can be seen in *Wouters*. Conduct of the NBA in regulating the legal profession was allocated for consideration under Article 81 EC—EC private law.[36] However, the specific rule promulgated by the association seemed more appropriately dealt with by EC public law.[37] The Court does not apply competition law norms, but instead subjects the functions to EC public law norms.[38]

2. Direct effect by the back door

The legitimacy of the teleological interpretation of the Treaty is not uncontested.[39] The imperative of market integration, and a view that a vigorous competition law was essential for its attainment, resulted in Article 81 EC operating as genuine supranational law.[40] Article 81 EC was immune from the interference by Member States first because competition law was not addressed to Member States and second, since there were few national competition regimes, no transfer of competence was involved.[41] However, this was not true of other policies necessary for the European project to succeed. Other policies required action by Member States, action that was

para 83, Joined Cases C-180/98 to C-184/98 *Pavel Pavlov v Stichting Pensioenfonds Medische Specialisten* [2000] ECR I-6451, para 87–88, and Cruz (2002: 148–149, 154).

[35] Gyselen (2000: 439) and references cites at note 38 below.

[36] Case C-309/99 *Wouters v Algemene Raad Van de Nederlandse Orde Van Advocaten (Raad Van de Balies Van de Europese Gemeenschap, Intervening)* [2002] ECR I-1577, para 86–96, AG Opinion para 116–133.

[37] Case C-309/99 *Wouters v Algemene Raad Van de Nederlandse Orde Van Advocaten (Raad Van de Balies Van de Europese Gemeenschap, Intervening)* [2002] ECR I-1577, para 32, 56, 100, AG Opinion para 29, 43, 84, 95, 113, 173–176 and Vossestein (2002: 842).

[38] Compare Case C-309/99 *Wouters v Algemene Raad Van de Nederlandse Orde Van Advocaten (Raad Van de Balies Van de Europese Gemeenschap, Intervening)* [2002] ECR I-1577, para 97, 109–110 with Case 120/1978 *Rewe-Zentral AG v Bundesmonopolverwaltung für Branntwein* [1979] ECR 649, para 8, Case 140/1979 *Chemial Farmaceutici SpA v DAF SpA* [1981] ECR 1, para 13, 16, Case C-415/93 *ASBL v Jean-Marc Bosman, Royal Club Liegeois SA v Jean-Marc Bosman and Others and UEFA v Jean-Marc Bosman* [1995] ECR I-4921, AG Opinion para 200, Barnard (2004: 54, n 55), Craig and de Búrca (2003: 617–626, 659), Cruz (2002: 91–104, 151–153), Forrester QC and Maclennan (2003: 550), Forrester and Norall (1984: 39–40), Monti (2002: 1081, 1086–1090), Mortelmans (2001: 637), P Oliver (1999: 804–806), O'Loughlin (2003: 68), Schmid (2000: 166–167), and Weatherill and Beaumont (1999: 472–474, 578–579).

[39] Craig and de Búrca (2003: 98–102).

[40] Wesseling (2000: 21, 32–33, 59–62), Wißmann (2000: 125–126), Cosma and Whish (2003: 52–53), and Gerber (1994: 103–107).

[41] Gerber (1994: 114–116) and Wesseling (2000: 63).

not forthcoming.[42] Koopmans asks, 'what can a court do when the political system fails to perform its functions?'[43] The Court's response to the period of stagnation and Member State inaction is the teleological approach: when the political imperative to achieve other Treaty aims stalled, the uncontested legitimacy of competition law was used as a Trojan horse to achieve non-efficiency (and on the thesis advanced, non-competition) aims indirectly.[44] The teleological interpretation avoids the need to implement various Community policies by applying them *inter alia* through the medium of Article 81 EC.

Article 2 EC and the various policy-linking clauses are incapable of direct effect.[45] In *Zaera*, the Court expressly rejects the idea that Article 2 EC can create rights for, or impose obligations on, individuals.[46] If a provision is incapable of direct effect, it is inappropriate to use another provision to indirectly create those same rights and obligations. It has been argued that the teleological approach, if justified, is justified in times of political stagnation or emergency.[47] This is not true of the present time; legitimate routes to achieve non-competition policy goals are open.[48] In the normalized environment, it may be that 'the Court is not prepared to continue "substituting" the judicial for the political process and is willing to accept the limits to the Community competences in policy-making'.[49] Though writing in a different context, the words of Advocate General Trabucchi are of relevance here; the interpretation of the Treaty must 'assume a significance to match the new situation which has come into being.'[50]

3. The intended relationship between Treaty provisions

A view of how Article 81 EC functions when solely concerned with efficiency is afforded by the UK Office of Fair Trading (OFT) assessment, made after close co-operation with the Commission, of collusion between

[42] Koopmans (1986: 927), Douglas-Scott (2002: 16–23), and Weiler (1999: 39).

[43] Koopmans (1986: 930).

[44] Wesseling (2000: 34–50, 61–64), Cini and McGowan (1998: 1, 12–13), and Weiler (1999: 39–63).

[45] On direct effect: Prinssen and Schrauwen (2002).

[46] Case 126/86 *Zaera v Instituto Nacional de la Seguridad Social* [1987] ECR 3697, para 11, Pescatore (1983: 162), and Prechal (2000: 1062).

[47] Hartley (1996: 102–109), Wesseling (2000: 63–64), Rasmussen (1986: 62–64), critiqued by Cappelletti (1987: text accompanying notes 2–3). Compare with Weiler (1999: 60–63).

[48] Monti (2002: 1092).

[49] Wesseling (2000: 68).

[50] Case 73/74 *Groupement des Fabricants de Papiers Peints de Belgique v Commission* [1975] ECR 1491, 1523.

British Midland and United Airlines in respect of transatlantic flights.[51] The agreement infringed Article 81(1) EC and the parties argued, and the OFT accepted, that Article 81(3) EC applied because scale and scope economies—increased quality, a greater number of flights on existing routes, and new routes—would result from the agreement.[52] Further, consumers would receive a fair share of the benefits in the form of lower fares (perhaps as great as 27 per cent), and a better quality of air travel.[53] The non-economic dis-benefits of an increased number of flights, particularly the environmental impact, were not considered.

Whilst efficiency is all that matters within Article 81 EC, efficiency is not all that matters. There are limits to what competition can achieve.[54] Whilst other Treaty goals cannot be balanced within Article 81 EC, those goals have not ceased to exist, nor have they become any less important. In the US context, *National Society of Professional Engineers v United States* considers how to address claims that competition law should be dis-applied in favour of other objectives in the public interest. The society argued that competition among professional engineers was contrary to the public interest. If engineers were allowed to compete on price, contracts would be awarded to the lowest bidder, regardless of quality, and this would be dangerous to the public health, safety, and welfare.[55] The society considered that preventing competition was 'a reasonable method of forestalling the public harm which might be produced by unrestrained competitive bidding'.[56] Rejecting the relevance of a public interest enquiry, Justice Stevens ruled:

> the purpose of the analysis is to form a judgment about the competitive significance of the restraint; it is not to decide whether a policy favoring competition is in the public interest . . . Subject to exceptions defined by statute, that policy decision has been made by the Congress.[57]

[51] Case CP/1535-01 *Notification by British Midland and United Airlines of Their Alliance Agreement* [2002] Decision of the Director General of Fair Trading 1 November, para 44 and Investigation of an Alliance Agreement in the Field of Air Transport [2001] OJ C 367/30.

[52] Case CP/1535-01 *Notification by British Midland and United Airlines of Their Alliance Agreement* [2002] Decision of the Director General of Fair Trading 1 November, para 134–140.

[53] Case CP/1535-01 *Notification by British Midland and United Airlines of Their Alliance Agreement* [2002] Decision of the Director General of Fair Trading 1 November, para 134–140.

[54] Deckert (2000: 176) and Monti (2002: 1089–1090).

[55] *National Society of Professional Engineers v United States* 435 US 679, 682–685 (1978).

[56] *National Society of Professional Engineers v United States* 435 US 679, 687 (1978).

[57] *National Society of Professional Engineers v United States* 435 US 679, 692 (1978). Also American Bar Association (Antitrust Section) and Hartley (chairman) (1999: 116–119). Pitofsky (1995) notes that this is perhaps an overly strong statement that does not reflect the true position.

The approach to conflicting goals suggested is that absent express derogation it is to be assumed that competition is in the public interest. The same approach is taken in the EC Treaty: 'where the Treaty intended to remove certain activities from the ambit of the competition rules, it made an express derogation to that effect.'[58] The circumstances in which Article 81 EC can be sacrificed for some other socially desirable goal must be expressly stated and clearly specified.[59] In enacting Article 86(2) EC there is recognition of 'a point to which the Treaty rules, particularly competition law, can be excluded'.[60] Article 36 EC and Article 296(1)(b) EC also provide express derogation, allowing Article 81 EC to be trumped in pursuit of other goals. If it were legitimate to balance competing goals within Article 81 EC these provisions would be otiose.

There may be conflict that is not resolved by an express provision in the Treaty. In such circumstances the Court has sometimes read the Treaty as allowing Member States to determine which public interests trump competition law.[61] However, the uniformity of Community law will not survive unilateral determinations that certain goals trump Article 81 EC.[62] The Court must decide which objectives override Article 81 EC and efficiency—a hierarchy of goals must be established.[63] The challenge of establishing a hierarchy is not a new one for the Court. The Community is an entity of limited competence, only able to act within the confines of powers conferred.[64] However, in exercising these conferred powers the Community is able to generate law capable of direct effect.[65] The doctrine of direct effect creates the possibility of conflict between norms derived from two sources; conflict between Community and

[58] Case T-61/89 *Dansk Pelsdyravlerforening v Commission* [1992] ECR-II 1931, para 54. Also Case 209/84 *Ministère Public v Asjes* [1986] ECR 1425, para 40–42.

[59] Goyder (1993: 4) and Van den Bergh and Camesasca (2001: 161). The UK Competition Act 1998, para 7 of Schedule 3, confers power to make such a determination on the Secretary of State.

[60] Auricchio (2001: 78). Also Weatherill and Beaumont (1999: 1005), Goyder (1998: 535), Faull and Nikpay (1999: para 5.122–5.155), and Louri (2002: 167–169).

[61] Case C-309/99 *Wouters v Algemene Raad Van de Nederlandse Orde Van Advocaten (Raad Van de Balies Van de Europese Gemeenschap, Intervening)* [2002] ECR I-1577, para 105, 108, 110, Auricchio (2001: 66–72). Schepel (2002: 38 n 27) writes that the public interest must be one contained within the Treaty.

[62] Schepel (2002: 38, 48).

[63] Kaysen and Turner (1959: 44–46), Panel Discussion (1998: 462), Kirchner (1998: 515), Monti (2002: 1069–1071, 1077–1078, 1094–1095), and Schmid (2000: 164–165).

[64] Article 5(1) EC, Dashwood (2004: 357–362), Lenaerts *et al.* (2005: 86–100), Craig and de Búrca (2003: 122–127), Dashwood (1998), and Dashwood (1996).

[65] Prinssen and Schrauwen (2002).

domestic law. The conflict is resolved by according Community law primacy.[66]

Whilst the Court has solved the problem of conflict between national and Community law, it has yet to clearly address the question of a hierarchy of norms within the Treaty itself.[67] Cruz argues that a hierarchy of goals can be identified in the Treaty: certain Treaty provisions presuppose the existence of others and those that are presupposed rank higher than those not presupposed; secondary legislation yields to directly effective Treaty provisions; Treaty provisions requiring implementing legislation yield to directly effective provisions.[68] This accords primacy to objectives over which competence has been fully transferred to the Community.[69] Wesseling on the other hand considers that a hierarchy of goals is lacking and that '[a]bsent a clear hierarchy...priorities are selected on a case by case basis'.[70] The ad hoc approach is uncertain, possibly to the extent of being contrary to the rule of law. Balancing within Article 81 EC only masks rather than resolves the problem. The problem is a constitutional rather than competition problem: it is at the constitutional level that a hierarchy of goals will ultimately be established.[71]

B. *Effectiveness of a unitary goal*

Whilst it is argued that the pursuit of multiple goals within Article 81 EC is illegitimate, the pursuit of multiple goals within Article 81 EC can also be seen to hinder the effectiveness of Article 81 EC in guiding conduct. Focusing on a unitary goal will enhance both legal certainty and the justiciability of the norm.

1. *Legal certainty*

Preventing the balancing of Treaty goals *within* Article 81 EC enhances legal certainty, transparency and accountability. Legal certainty is described by Hartley as 'one of the most important general principles recognised by the European Court'.[72] Though not easily explained, 'predictability is probably

[66] Case 6/64 *Flaminio Costa v ENEL* [1964] ECR 585, Case 106/77 *Amministrazione Delle Finanze Dello Stato v Simmenthal SpA* [1978] ECR 629, Alter (2001), Craig and de Búrca (2003: 275–316), Weatherill and Beaumont (1999: 433–453), and Weiler (2001: 55–57).

[67] This problem is considered in Ward (2003), Cruz (2002), and Bieber and Salomé (1996).

[68] Cruz (2002: 63–66, 69–72, 76–80). [69] Cruz (2002: 63–66).

[70] Wesseling (2000: 48–49). [71] Bork (1993: 72–89) and Kovacic (1990: 1438–1440).

[72] Hartley (1998: 142). Also Case C-234/89 *Stergios Delimitis v Henninger Bräu AG* [1991] ECR I-935, para 47 and Arnull *et al.* (2000: 137).

the core aspect of it'.[73] Article 81 EC is intended to guide those who produce or consume as to what they can expect from the marketplace. The obligation imposed is not to collude to contrive a scarcity of output, which results in allocative inefficiency (Chapter 5). Undertakings may derogate from this obligation in pursuit of productive efficiency (Chapter 6). The obligation and the derogation are related: there is an obligation not to cause inefficiency and a derogation to allow pursuit of efficiency. It seems that there is no obligation to confer exemption if the Article 81(3) EC conditions are met.[74] Some take the view that although the Article 81(3) EC conditions are satisfied, the provision may not be applied unless, in addition, collusion promotes economic and social justice or furthers an environmental, energy, employment or structural goal.[75] However, there must be congruence between the law as declared and the law as administered.[76] When other goals are considered the express conditions of Article 81 EC cease to be determinative and the link between the obligation and the derogation is broken. Those with rights under Article 81(1) EC do not anticipate that 81(3) EC will depend on factors unrelated to the infringement of their Article 81(1) EC rights. Those entitled to have Article 81(3) EC exercised in their favour do not expect recourse to the provision to be denied because of factors unrelated to their obligations under Article 81(1) EC. To some extent guidelines and soft law instruments can be used to reduce uncertainty.[77] However, unless the law is certain it cannot guide conduct; uncertain law is ineffective.[78] One must be mindful that '[s]acrificing too many social goals on the altar of [legal certainty] may make the law barren and empty'.[79] It may well be the concern with broader social issues that tips the social consensus towards support for competition law.[80] No claim is made that the various goals should not be pursued; the claim is that to pursue them *within* Article 81 EC is detrimental to legal certainty.[81]

[73] Hartley (1998: 143).

[74] Cosma and Whish (2003: 42–45), Ritter *et al.* (2000: 115), Temple Lang (1981: 344), and Ehlermann (1996: 93). *Contra* Deringer (1968: para 405, 407–409) and Guidelines on the Application of Article 81(3) of the Treaty [2004] OJ C 101/97, para 42.

[75] Bouterse (1994: 21, 32–37), Faull and Nikpay (1999: para 2.219–2.131, 2.144–2.145), Ritter *et al.* (2000: 122–124), and Mortelmans (2001: 638–642). Compare with Monti (2002: 1069–1071, 1077–1078). *Contra* Deringer (1968: para 405, 407–409) and Wils (2002b: §1.1.3).

[76] Fuller (1969: 81–91).

[77] Cosma and Whish (2003: 42).

[78] Fuller (1969: 63–65, 155–157) and Cini and McGowan (1998: 28).

[79] Raz (1979: 229).

[80] Scherer and Ross (1990: 18–19), Fels and Edwards (1998: 58), Neven (1998: 111), and Schaub (1998: 125–126).

[81] Neven *et al.* (1998: 18–19) and Castañeda (1998: 51).

Legal certainty is also linked to the corollary principle that 'those who act in good faith on the basis of the law as it is or as it seems to be should not be frustrated in their expectations'.[82] Further, in order to evaluate the effectiveness of competition law in the attainment of its aims it is necessary to have a clear, unitary aim.[83] Finally, the forces of globalization have resulted in pressures to converge on a standard norm, which makes it possible to comply with the law in multiple jurisdictions, enhancing legal certainty, and minimizing the cost of compliance.[84] Eliminating non-efficiency factors will enhance legal certainty; though not make the law certain. Amato acknowledges that it would be naive to expect that if multiple goals are removed then:

> the uncertainties will disappear too: without diverse options to choose among due to the political and administrative discretion in pursuing manifold public policies, there will be only the single, true solution required in each case by antitrust rules finally cleansed of all the rest.[85]

2. *Justiciability*

Balancing competing goals *within* Article 81 EC requires the decision maker to decide how much competition to sacrifice in favour of a series of vaguely specified goals.[86] This approach is described as 'essentially a political act, whose subject matter is the weighing up of incommensurable factors'.[87] Asked how courts can balance conflicting goals when applying competition law Judge Jonson, of the Swedish Market Court, replied '[m]y first answer would be that nobody really can'.[88] The reason for adopting Article 9(1) of Regulation 17/62 was that Article 81(3) EC involved so much discretion that a centralized system of enforcement was the only way to ensure a consistent interpretation.[89] The Monopolkommission is adamant that 'the justiciability

[82] Arnull *et al.* (2000: 137). On legitimate expectation: Craig and de Búrca (2003: 380–387) and Hartley (1998: 145–147).

[83] Furse (1996: 257–258), Pitofsky (1995: 25), Rossi and Wright (1984), Gordon and Morse (1975), and Pawson and Tilley (1997: 1–29).

[84] Calvani (2003), Ehlermann and Laudati (1998: vi-xiii, 347), and Amato (1997: 124–129).

[85] Amato (1997: 123). Also Castañeda (1998: 50).

[86] Furse (1996: 255), Bovis (2001: 102), Lever QC and Peretz (1999: para 3.2), Kingston (2001: 348), Holmes (2000: 67) and Wesseling (2001: 368–374).

[87] Monopolkommission (2000: 37). Also Bouterse (1994: 36, 44), Cooke (2000: 61–63, 66–67), Monopolkommission (2000: 38), Panel Discussion (1998: 485), Pescatore (1983: 174–175), Schmid (2000: 156, 166), Wesseling (1999: 424–425), Wesseling (2000: 18), Whish and Sufrin (2000: 150), and Whish (2000a: 75–78).

[88] Organisation for Economic Co-operation and Development (1997: 88).

[89] Bouterse (1994: 30), White Paper on Modernisation of the Rules Implementing Articles 85 and 86 of the EC Treaty Commission Programme No 99/027 [1999] OJ C 132/1, para 17, 77,

of norms of Community law is an essential requirement for their direct applicability'.[90] It has been argued that national courts and competition authorities are ill placed to balance a broad range of competing Community policies; a justiciable norm requires a unitary goal.[91]

IV. Conclusion

Chapters 5 and 6 have argued that, as a substantive matter, Article 81 EC is concerned with efficiency. Article 81(1) EC is infringed by collusion with the aim or consequence of reduced output and increased price—allocative inefficiency. Article 81(3) EC allows Article 81(1) EC to be declared inapplicable when the collusion also results in productive efficiency, so that net efficiencies are considered. An efficiency-only role for Article 81 EC is defended on the grounds that (a) considering non-efficiency goals confers rights and imposes obligations on those not properly entitled to those rights or subject to those obligations by eroding the public/private divide; (b) it is illegitimate to use Article 81 EC as a Trojan horse to impose obligations or deny rights to private parties on the basis of Treaty provisions incapable of horizontal direct effect; (c) balancing competing goals within Article 81 EC renders otiose Treaty provisions resolving conflicts of values; (d) a unitary goal provides greater legal certainty; and (e) a unitary goal is more justiciable than a norm pursuing multiple goals. Hard cases in Community competition law are those in which efficiency and other values cannot be pursued simultaneously, cases that reveal tension between Treaty provisions. They are cases when non-efficiency values seek to triumph, cases in which reflection is had as to whether efficiency is desirable at all costs. It is acknowledged that though efficiency is all that matters in

and Monopolkommission (2000: 20). The non-justiciability of Article 81(3) EC is said to be illustrated by the Court of Justice's limited review of its application: Wesseling (2001: 371–372). *Contra* Ehlermann and Atanasiu (2002: 74–75). On the justiciability of competition issues in general: Stevens and Yamey (1965: 32–49).

[90] Monopolkommission (2000: 19). Also Case 106/77 *Amministrazione Delle Finanze Dello Stato v Simmenthal SpA* [1978] ECR 629, para 14, Case 26/62 *Van Gend en Loos v Netherlands Inland Revenue Administration* [1963] ECR 1, AG Opinion 22–24, Craig and de Búrca (1998: 168), Prechal (2000: 1062), and Wesseling (2000: 56).

[91] Whish (2003: 154–156), Venit (2003: 578–579), Yarrow (2002: 2–3), Schaub (1998: 125–126), Neven (1998: 113, 116–117), and Morris (2002: 15–18). Gyselen (2002: 188–197) argues that balancing various Treaty goals does not affect justiciability.

Article 81 EC, efficiency is not all that matters.[92] However, the balancing of Article 81 EC against other Treaty goals is properly external to the competition law assessment. The Treaty contains express provisions that structure an enquiry into whether other Treaty goals trump Article 81 EC. Absent express provision, the situation is more uncertain. Though the Court has resolved the issue of conflict between national and Community norms, it has done little to resolve issues of conflict between Treaty norms. The problem is a constitutional rather than competition problem and it is at the constitutional level that resolution is required.

[92] Deckert (2000: 176) and Monti (2002: 1089–1090).

8

The boundaries of Article 81 clarified?

This book, setting out from a position of uncertainty over what Article 81 EC seeks to achieve and how it operates to achieve it, attempts to identify the normative boundaries of Article 81 EC. The initial problem is that what competition is, and hence what a restriction entails, is undefined. There is a question over whether competition is valued to promote efficiency, to achieve the Community objective of market integration, or to promote certain market freedoms desirable in a democracy. Chapter 2 recognized that these values may be achieved by Article 81 EC, and may be achieved simultaneously, but argued that, when the values conflict, Article 81 EC increasingly ranks the ability to make the most of society's scarce resources as superior. It can thus be said that the function of Article 81 EC is to protect and promote efficiency and that a restriction of competition entails causing inefficiency.

The nature of the obligation must be an appropriate one to impose on those bound by the provision. Chapter 3 supports the conception of the Article 81 EC obligation by considering the identity of the addressees of the provision. The Court takes a functional approach to identify the addressees of the obligation; the functional approach is used to exclude from the reach of Article 81 EC action taken to meet obligations arising in a social democracy that require the use of public power to fulfil; the exercise of such power requires distinctive substantive, justificatory, and procedural rules, and as such lies beyond the scope of Article 81 EC. The functional approach instead captures actors with the potential to make profit by taking the financial risk of offering goods or services to the market. Article 81 EC is clearly seen as an aspect of private law.[1] Jurisdiction rests on a public/private distinction and the substantive and justificatory rules are cast to generate obligations appropriate for private actors to perform.

[1] Whish (2003: 211).

Entities engaged in economic activity, exercising private power, are appropriately subject to the obligation not to cause inefficiency. The obligation is not heroic; the obligation is narrowed because it is collusion that must not be used to cause inefficiency.[2] The collusive concepts of agreement and concerted practices are considered in Chapter 4. Both agreement and concerted practice are seen as methods by which undertakings can reduce uncertainty about the future conduct of others. This conception identifies conduct appropriate for scrutiny under Article 81 EC, as it is uncertainty as to the future that forces undertakings to engage in a competitive struggle. The collusive concepts should be seen to serve a jurisdictional rather than substantive function; this makes it possible to understand and tolerate their broad scope. The existence of collusion is separate from the question of whether the collusion causes harm to competition.

Article 81 EC involves a bifurcated analysis. Article 81(1) EC contains a prohibition; Article 81(3) EC allows the prohibition to be dis-applied if certain conditions are met. What is unclear is 'the division of labour between paragraphs (1) and (3) of Article [81]'.[3] Whish asks whether 'the bifurcation of Article [81 is] the "original sin" of European competition law?'[4] It is felt that the bifurcation of Article 81 EC 'grossly complicates, with little or no redeeming virtues, both the formulation and enforcement of specific rules'.[5] The Article 81(1) EC prohibition is the subject of Chapter 5. Restriction of competition in Article 81(1) EC serves a substantive function; the substantive obligation imposed on those engaged in economic activity is not to collude to cause, or engage in collusion causing, allocative efficiency. A classic critique of Article 81(1) EC is, particularly when applied by the Commission, that this economic effect is not established. However, allocative inefficiency is shown to be central by examining the factors causing the Court to proclaim that restricted competition exists. Allocative efficiency can be measured, predicted, or presumed to exist. Whether the obligation not to collude to cause allocative inefficiency has been breached can be demonstrated *ex post* by measurement. Measurement of market output and price can only be carried out in an *ex post* analysis, ie, after the claimed anti-competitive consequences have occurred. A system of *ex ante* control requires tools to predict whether anti-competitive outcomes are more likely than not. Whether the obligation not to collude to cause

[2] Case T-41/96 *Bayer AG v Commission* [2000] ECR II-3383, para 64.
[3] Forrester and Norall (1984: 20). [4] Panel Discussion (1998: 462).
[5] Hawk (1988: 54).

allocative inefficiency is likely to be breached can be demonstrated *ex ante* with economic predictive tools. Allocative efficiency demonstrated by measurement or prediction is said to have the *effect* of restricting competition. Legal presumptions that allocative inefficiency exists are also used. Allocative inefficiency established by legal presumption is said to have the *object* of restricting competition. The extent to which it is necessary to show the *effect* of restricting competition is determined by the scope and content of what is said to have the *object* of restricting competition. The wider the range of circumstances covered by legal presumptions, the narrower the range of circumstances in which it is necessary to show the effect on competition; enforcement activity appears focused on *object* cases.[6] Measurement and prediction are avoided, and legitimately so, if a presumption of allocative inefficiency exists. Rather than acting *ultra vires*, the Commission displays a bias for one of two legal bases for a prohibition decision—for object over effect.[7] Great attention must thus be placed on the content of the legal presumptions and how they are established.

Allocative inefficiency is too narrow a conception of the competition law problem. Attention is also paid to whether conduct increases, is neutral, or reduces productive efficiency, which has a greater impact on living standards. Porter argues that all 'practices scrutinized in antitrust should be subjected to the following question: How will they affect productivity growth?'[8] Chapter 6 shows Article 81(3) EC as providing a framework in which to assess an undertaking's claim that the loss in allocative inefficiency identified under Article 81(1) EC is offset by a gain in productive efficiency. The distribution of identified productive efficiency gains is and ought to be tilted in favour of consumers, as it is consumers that forgo allocative efficiency to generate the productive efficiency gains. In arguing that Article 81(3) involves a productive efficiency enquiry, the coherence and underlying logic of the bifurcated analysis contained in Article 81 is revealed and answers those that consider it illogical to exempt collusion prohibited as anti-competitive under Article 81(1) EC because it promotes competition under Article 81(3) EC.[9]

[6] White Paper on Modernisation of the Rules Implementing Articles 85 and 86 of the EC Treaty Commission Programme No 99/027 [1999] OJ C 132/1, executive summary para 13, third recital of Council Regulation (EC) No 1/2003 [2003] OJ L 1/1, Gerber (2001: 56), and Venit (2003: 546).

[7] The same phenomena can be observed in US antitrust: Arthur (2000: 349–350) and Gavil (2000b: 331–332).

[8] Porter (2001: 932).

[9] Fox (2001: 142 note 6), Wesseling (2001: 359), and Monti (2002: 1060–1061).

With Article 81(1) EC as an allocative efficiency enquiry, Article 81(3) EC as a productive efficiency enquiry, and belated recognition of the view of Advocate General Lagrange that 'it would be contrary to the most elementary considerations of justice to allow the application of the prohibition contained in Article [81(1)] . . . without allowing the enterprises concerned to avail themselves of the provisions of Article [81(3)]', it might be questioned whether bifurcation still matters.[10] It remains important to remember that '[w]hether an agreement is caught by Article [81(1)] and whether it benefits from the exemption in Article [81(3)] are questions which do not depend on the same conditions or have the same consequences'.[11] This is so if only because the burden of proof is allocated differently under Article 81(1) and 81(3) EC.[12] In order to know who must prove what, Article 81(1) EC issues cannot and ought not to be commingled with Article 81(3) EC issues.[13] Further, the distinction remains vitally important because the circumstances to which Article 81(1) and 81(3) EC are being applied can 'be expected to be in a state of continuous change'.[14] And Regulation 1/2003 contains no equivalent of Article 8(1) of Regulation 17, which provides that '[a] decision in application of Article [81(3)] of the Treaty shall be *issued for a specified period*'.[15] When market factors change, it will be important to know how they affect the assessment made. For example, Articles 28 and 29 of Regulation 1/2003 and Articles 6 and 7 of Regulation 2790/1999 allow the benefit of a block exemption to be withdrawn, whilst Article 8 of Regulation 2790/1999 allows the Commission to prevent whole sectors from benefiting from the regulation.[16] An actual infringement of Article 81(1) EC must be shown

[10] Case 13/61 *Kledingverkoopbedrijf de Geus en Untdenbogerd v Robert Bosch GmbH* [1962] ECR 45, AG Opinion 66. Also Article 1(2) of Council Regulation (EC) No 1/2003 [2003] OJ L 1/1.

[11] Case 32/65 *Italian Republic v Council of the EEC and Commission of the EEC* [1966] ECR 389, 405. Also Case 56/65 *Société Technique Minière v Maschinenbau Ulm GmbH* [1966] ECR 235, AG Opinion 259, Case T-17/1993 *Matra Hachette SA v Commission* [1994] ECR II-595, para 48, and Joined cases T-374/94, T-375/94, T-384/94, T-388/94 *European Night Services v Commission* [1998] ECR II-3141, para 81, 90–160, 206, 229.

[12] Article 2 of Council Regulation (EC) No 1/2003 [2003] OJ L 1/1 and Wesseling (2001: 359).

[13] On the importance of a clear allocation of burdens: Collins (1999: 60).

[14] *David John Passmore v Morland* [1999] European Commercial Cases 367, para 21.

[15] Regulation 17. First Regulation Implementing Articles 81 and 82 of the Treaty [1962] OJ Special Edition 204/62. Also Bourgeois and Humpe (2002: 50).

[16] Commission Regulation (EC) No 2790/1999 [1999] OJ L 336/21, recital 13 and Articles 6–7 of Commission Regulation (EC) No 2790/1999 [1999] OJ L 336/21, Guidelines on Vertical Restraints [2000] OJ C 291/1, para 71–87, Whish (2000b: 918–921), Subiotto and Amato (2000: 15), and Subiotto and Amato (2001: 183–185). On the situation under Regulation 17: Korah (2000: para 2.5.2.2) and Roth QC (2001: para 3–084).

before withdrawal and dis-application of an exemption can have any consequences; not all those in the sector will be party to collusion infringing Article 81(1) EC and a clear distinction between Article 81(1) and 81(3) EC is required to determine which undertakings are affected.

With Article 81(1) EC as an allocative efficiency enquiry and Article 81(3) EC as a productive efficiency enquiry, Article 81 EC as a whole can operate and be understood as operating to consider net efficiencies. Chapter 7 advances efficiency as the exclusive concern. It is acknowledged that efficiency is 'weak because it is only one of many values considered important by our society'.[17] It is also acknowledged that non-efficiency goals have, in the past, been considered in Article 81 EC. However, legal certainty, justiciability, the legitimacy of a teleological interpretation, and the relationship between Article 81 EC and other Treaty provisions, show that the extent to which non-efficiency goals trump efficiency is a question external to Article 81 EC. The hard cases are those in which non-efficiency values seek to triumph, cases in which reflection is had as to whether efficiency is to be had at all costs. It is acknowledged that efficiency is not all that matters.[18] It is simply all that matters in Article 81 EC, particularly since the Treaty contains express provisions which structure an enquiry into whether other Treaty goals trump Article 81 EC.

The argument has been that after close to 50 years of application, a green paper, an action plan, a new block exemption regulation, guidelines, a white paper, observations, a new procedural regulation, and more notices and guidelines–a list which does not include all that has occurred–it is possible to understand Article 81 EC and what it seeks to achieve.[19] The aim has been to reveal the intellectual order and rational structure underlying Article 81 EC and in doing so provide normative justification for the substantive features of Article 81 EC. It has been argued that actors and activities falling within the scope of Article 81 EC are subject to a substantive obligation prohibiting contrived reductions in output. Since

[17] Areeda and Kaplow (1997: 6). [18] Deckert (2000: 176) and Monti (2002: 1089–1090).

[19] Green Paper on Vertical Restraints in EC Competition Policy Com (96) 721 Final [1997] 4 CMLR 519, para 7, 219–269, Communication from the Commission on the Application of the Community Competition Rules to Vertical Restraints (Follow-up to the Green Paper on Vertical Restraints) Com(98)544 Final [1998] OJ C 365/3, Commission Regulation (EC) No 2790/1999 [1999] OJ L 336/21, Guidelines on Vertical Restraints [2000] OJ C291/1, White Paper on Modernisation of the Rules Implementing Articles 85 and 86 of the EC Treaty Commission Programme No 99/027 [1999] OJ C 132/1, White Paper on Reform of Regulation 17: Summary of the Observations [2001] 4 CMLR 10, Council Regulation (EC) No 1/2003 [2003] OJ L 1/1, and Commission Press Release IP/04/411 30.03.2004.

output reduction can co-exist with cost reduction and innovation, and these latter features are desirable, cost reduction and innovation operate to justify infringement of the substantive obligation. The thesis is that output, cost and innovation are the only legitimate issues in an Article 81 EC analysis. It would be naive to think that this result makes the boundaries of Article 81 EC clear. However, it is hoped that a contribution to making the boundaries less opaque has been made.

BIBLIOGRAPHY

ABBAMONTE, GIUSEPPE B (1998) 'Cross-Subsidisation and Community Competition Rules: Efficient Pricing Versus Equity?' 23 *ELRev* 414–433.

ADAMS, WALTER (1953) 'Competition, Monopoly and Countervailing Power' 67 *Quart J Econ* 469–492.

——(1991) 'The Case for Structural Tests' in Brock, James W and Kenneth G Elzinga (eds) *Antitrust, the Market, and the State: The Contributions of Walter Adams* (Armonk, New York: M E Sharpe).

ADELMAN, M A (1949) 'Integration and Antitrust Policy' 63 *Harv L Rev* 27–77.

AGHION, PHILIPPE, and PATRICK BOLTON (1987) 'Contracts as a Barrier to Entry' 77 *Amer Econ Rev* 388–401.

—— and PETER HOWITT (1992) 'A Model of Growth through Creative Destruction' 60 *Econometrica* 323–351.

AHDAR, REX (2002) 'Consumers, Redistribution of Income and the Purpose of Competition Law' 23 *ECLR* 341–353.

ALBORS-LLORENS, ALBERTINA (1999) 'The European Court of Justice, More Than A Teleological Court' 2 *CYELS* 373–398.

——(2002) 'Competition Policy and the Shaping of the Single Market' in Barnard, Catherine and Joanne Scott (eds) *The Law of the Single European Market* (Oxford: Hart) 311–331.

ALDER, JOHN (1997) 'Obsolescence and Renewal: Judicial Review in the Private Sector' in Leyland, Peter and Terry Woods (eds) *Administrative Law Facing the Future: Old Constraints and New Horizons* (London: Blackstone) 160–183.

ALLISON, JOHN W F (1997) 'Theoretical and Institutional Underpinnings of a Separate Administrative Law' in Taggart, Michael (ed) *The Province of Administrative Law* (Oxford: Hart) 71–89.

ALTER, KAREN J (2001) *Establishing the Supremacy of European Law: The Making of an International Rule of Law in Europe* (Oxford: Oxford University Press).

AMATO, GUILIANO (1997) *Antitrust and the Bounds of Power: The Dilemma of Liberal Democracy in the History of the Market* (Oxford: Hart).

American Bar Association (Antitrust Section), and James E Hartley (chairman) (1999) *Monograph No. 23: The Rule of Reason* (Chicago: ABA Section of Antitrust Law).

AMRAMOVITZ, MOSES (1989) 'Economic Goals and Social Welfare in the Next Generation' in Amramovitz, Moses (ed) *Thinking About Growth and Other Essays on Economic Growth and Welfare* (Cambridge: Cambridge University Press) 301–307.

ANDREWS, PHILIP (2002) 'Self-Regulation by Professions—the Approach under EU and US Competition Rules' 23 *ECLR* 281–285.

ANLAUF, TODD J (2000) 'Severing Ties with the Strained Per Se Test for Antitrust Tying Liability: The Economic and Legal Rationale for A Rule of Reason' 23 *Hamline L Rev* 476–510.

ANTHONY, ROBERT A (1992) 'Interpretive Rules, Policy Statements, Guidelines, Manuals, and the Like: Should Federal Agencies Use Them to Bind the Public?' 41 *Duke LJ* 1311–1384.

ANTUNES, LUIS MIGUEL PAIS (1991) 'Agreements and Concerted Practices under EEC Competition Law: Is the Distinction Relevant?' 11 *YEL* 57–77.

AREEDA, PHILLIP (1983) 'Market Definition and Horizontal Restraints' 52 *Antitrust LJ* 553–585.

——(1986) 'The Rule of Reason—A Catechism on Competition' 55 *Antitrust LJ* 571–589.

—— and LOUIS KAPLOW (1997) *Antitrust Analysis: Problems, Text, Cases* (Fifth Edition New York: Aspen Law & Business).

ARNULL, ANTHONY M, ALAN A DASHWOOD, MALCOLM G ROSS, and DERRICK A WYATT (2000) *Wyatt & Dashwood's European Union Law* (Fourth Edition London: Sweet & Maxwell).

ARONSON, MARK (1997) 'A Public Lawyer's Responses to Privatisation and Outsourcing' in Taggart, Michael (ed) *The Province of Administrative Law* (Oxford: Hart) 40–70.

ARTHUR, THOMAS C (1994) 'The Costly Quest for Perfect Competition: Kodak and Nonstructural Market Power' 69 *NYU L Rev* 1–76.

——(2000) 'A Workable Rule of Reason: A Less Ambitious Antitrust Role for the Federal Courts' 68 *Antitrust LJ* 337–389.

ASCH, PETER (1970, 1984 reprint) *Economic Theory and the Antitrust Dilemma* (Malabar, Florida: Robert E Krieger Publishing).

AURICCHIO, VITO (2001) 'Services of General Economic Interest and the Application of EC Competition Law' 24 *W Comp* 65–91.

BAARDMAN, B (1971) 'Annotation' 8 *CML Rev* 89–92.

BADEN FULLER, C W F (1981) 'Economic Issues Relating to Property Rights in Trademarks: Export Bans, Differential Pricing, Restrictions on Resale & Repackaging' 6 *ELRev* 162–179.

BAILEY, DAVID (2004) 'Scope of Judicial Review under Article 81 EC' 41 *CML Rev* 1327–1360.

BAIN, JOE STATEN (1993. Originally published Harvard University Press, Cambridge, 1956) *Barriers to New Competition: Their Character and Consequences in Manufacturing Industries* (Fairfield, New Jersey: Augustus M Kelley).

BAINBRIDGE, DAVID (1999) *Intellectual Property* (Fourth Edition London: Pitman).

BAKER, JONATHAN B, and TIMOTHY F BRESNAHAN (1992) 'Empirical Methods of Identifying and Measuring Market Power' 61 *Antitrust LJ* 3–16.

——(1996) 'Vertical Restraints with Horizontal Consequences: Competitive Effects Of "Most-Favored-Customer Clauses"' 64 *Antitrust LJ* 517–534.

BAKER, SIMON, and LAWRENCE WU (1998) 'Applying the Market Definition Guidelines of the European Commission' 19 *ECLR* 273–280.

BAMFORTH, NICHOLAS (1997) 'The Public Law—Private Law Distinction: A Comparative and Philosophical Approach' in Leyland, Peter and Terry Woods (eds) *Administrative Law Facing the Future: Old Constraints and New Horizons* (London: Blackstone) 136–159.

BANKS, DAVID (1997) 'Non-Competition Factors and Their Future Relevance under European Merger Law' 18 *ECLR* 182–186.

BARNARD, CATHERINE (2004) *The Substantive Law of the EU: The Four Freedoms* (Oxford: Oxford University Press).

BARRY, NORMAN P (1989) 'Political and Economic Thought of German Neo-Liberals' in Peacock, Alan and Hans Willgerodt (eds) *German Neo-Liberals and the Social Market Economy* (London: Macmillan).

BATOR, FRANCIS M (1957) 'The Simple Analytics of Welfare Maximization' 47 *Amer Econ Rev* 22–59.

——(1960) *The Question of Government Spending: Public Needs and Private Wants* (New York: Harper).

BAYLES, MICHAEL D (1987) *Principles of Law: A Normative Analysis* (Dordrecht: D Reidel).

BEALE, HUGH (2004) *Chitty on Contracts* (Twenty-ninth edition London: Sweet & Maxwell).

BEATSON, J (2002) *Anson's Law of Contract* (Twenty-eighth edition Oxford: Oxford University Press).

BEGG, DAVID, STANLEY FISCHER, and RUDIGER DORNBUSCH (1997) *Economics* (Fifth Edition Maidenhead: McGraw-Hill).

——RUDIGER DORNBUSCH, and STANLEY FISCHER (2000) *Economics* (Sixth Edition London: McGraw-Hill).

BELHAJ, SOMAYA, and JOHAN W VAN DE DRONDEN (2004) 'Some Room for Competition Does Not Make A Sickness Fund an Undertaking. Is EC Competition Law Applicable to the Heath Care Sector? Joined Cases C-264/01, C-306/01, C-453/01 and C-355/01 AOK' 11 *ECLR* 682–687.

BELOFF, MICHAEL J (1989) 'Pitch, Pool, Rink,... Court? Judicial Review in the Sporting World' *Public Law* 95–110.

BENGOETXEA, JOXERRAMON, NEIL MACCORMICK, and LEORNOR MORAL SORIANO (2001) 'The European Court of Justice' in De Búrca, Gràinne and J H H Weiler (eds) *Integration and Integrity in the Legal Reasoning of the European Court of Justice* (Oxford: Oxford University Press) 43–85.

BESCHLE, DONALD L (1987) 'What, Never? Well, Hardly Ever: Strict Antitrust Scrutiny as an Alternative to Per Se Antitrust Illegality' 38 *Hastings LJ* 471–515.

BEUTEL, PHILLIP A (2002) 'The Intersection of Antitrust and Intellectual Property Economics: A Schumpeterian View' November/December *Antitrust Insights: NERA Newsletter* 1–7.

BIEBER, ROLAND, and ISABELLE SALOMÉ (1996) 'Hierarchy of Norms in European Law' 33 *CML Rev* 907–930.

BISHOP, BILL, and SIMON BISHOP (1996) 'Reforming Competition Policy: Bundeskartellamt—Model or Muddle' 17 *ECLR* 207–209.

BISHOP, BILL (2000) 'The Anti-Trust Epidemic—Causes and Prospects' in Rivas, José and Margot Horspool (eds) *Modernisation and Decentralisation of EC Competition Law* (The Hague: Kluwer Law International) 39–47.

BISHOP, SIMON, and MIKE WALKER (1999) *Economics of EC Competition Law: Concepts, Application and Measurement* (London: Sweet & Maxwell).

—— and DEREK RIDYARD (2002) 'EC Vertical Restraints Guidelines: Effects-Based or Per Se Policy?' 23 *ECLR* 35–37.

——(2003) 'Pro-Competitive Exclusive Supply Agreements: How Refreshing!' 24 *ECLR* 229–232.

BLACK, OLIVER (1992) 'Communication and Obligation in Arrangements and Concerted Practices' 13 *ECLR* 200–205.

——(1997) 'Per Se Rules and Rules of Reason: What Are They' 18 *ECLR* 145–161.

——(2003a) 'Joint Action, Reliance and the Law' 14 *Kings College Law Journal* 65–80.

——(2003b) 'Collusion and Co-Ordination in EC Merger Control' 24 *ECLR* 408–411.

——(2003c) 'Grades of Correlation: The Spectrum from Independent Action to Collusion, and Its Implications for Antitrust' 2 *Competition Law Journal* 102–109.

——(2003d) 'What Is an Agreement?' 24 *ECLR* 504–509.

——(2004a) 'Reliance and Obligation' 17 *Ratio Juris* 269–284.

——(2004b) 'Agreements, Undertakings, and Practical Reason' 10 *Legal Theory* 77–95.

——(2005) *Conceptual Foundations of Antitrust* (Cambridge: Cambridge University Press).

BLAIR, ROGER D, and DAVID L KASERMAN (1985) *Antitrust Economics* (Homewood, Illinois: Richard D Irwin).

BLAUG, MARK (2001) 'Is Competition Such A Good Thing? Static Efficiency Versus Dynamic Efficiency' 19 *Rev Ind Organ* 37–48.

BODNER Jr, JOHN (1988) 'The Tunney Act 'Public Interest' Provision: A New Weapon in Hostile Takeover Battles' 3 *Antitrust* 20–23.

BÖHM, FRANZ (1989) 'Rule of Law in A Market Economy' in Peacock, Alan and Hans Willgerodt (eds) *Germany's Social Market Economy: Origins and Evolution* (Basingstoke: Macmillan).

BONI, STEFANO, and PIETRO MANZINI (2001) 'National Social Legislation and EC Antitrust Law' 24 *W Comp* 239–256.

BORK, R H (1965) 'The Rule of Reason and the Per Se Concept: Price Fixing and Market Division, I' 74 *Yale LJ* 775–847.

BORK, ROBERT H (1967a) 'Antitrust and Monopoly: The Goals of Antitrust Policy' 57 *Amer Econ Rev* 242–253.

——(1967b) 'A Reply to Professors Gould and Yamey' 76 *Yale LJ* 731–743.

——(1993) *The Antitrust Paradox: A Policy at War with Itself* (New York: Free Press).

BORRIE, SIR GORDON (1989) 'The Regulation of Public and Private Power' *Public Law* 552–567.

Boscheck, Ralf (2000) 'The EU Policy Reform on Vertical Restraints—an Economic Perspective' 23 *W Comp* 3–49.

Boulding, K E (1945) 'In Defense of Monopoly' 59 *Quart J Econ* 524–542.

Bourgeois, Jacques H J, and Christophe Humpe (2002) 'The Commission's Draft "New Regulation 17"' 23 *ECLR* 43–51.

Bouterse, R B (1994) *Competition and Integration: What Goals Count? EEC Competition Law and Goals of Industrial, Monetary, and Cultural Policy* (Deventer: Kluwer Law and Taxation).

Bovis, Christopher (2001) 'Transforming the Application of EC Competition Laws: The Case of Decentralisation' 12 *EBLR* 98–104.

Bowles, Samuel, and Herbert Gintis (1993) 'The Revenge of Homo Economicus: Contested Exchange and the Revival of Political Economy' 7 *J Econ Perspect* 83–102.

Brealey, Mark (1985) 'The Burden of Proof before the European Court' 10 *ELRev* 250–262.

Brennan, Timothy J (2001) 'Do Easy Cases Make Bad Law? Antitrust Innovations or Missed Opportunities in United States v Microsoft' 69 *Geo Wash L Rev* 1042–1102.

Bright, Chris, and Kate Currie (2003) 'Is Bettercare a Better Pill' 24 *ECLR* 41–45.

Bright, Christopher (1995) 'Deregulation of EC Competition Policy: Rethinking Article 85(1)' in Hawk, Barry E (ed) *Annual Proceedings of the Fordham Corporate Law Institute: International Antitrust Law & Policy 1994* (Irvington-on-Hudson, NY: Transnational Juris Publications).

——(1996) 'EU Competition Policy: Rules, Objectives and Deregulation' 16 *OJLS* 535–560.

Broder, Douglas F (2005) *A Guide to US Antitrust Law* (London: Sweet & Maxwell).

Brown, Lionel Neville, and Tom Kennedy (2000) *The Court of Justice of the European Communities* (Fifth Edition London: Sweet & Maxwell).

Brunn, Niklas, and Jari hellsten, eds. (2001) *Collective Agreement and Competition Law in the EU: The Report of the COLCOM-Project* (Copenhagen: DJF).

Buckley, F H (2005) *Just Exchange: A Theory of Contracts* (London: Routledge).

Buendia Sierra, José Luis (1999) *Exclusive Rights and State Monopolies under EC Law: Article 86 (Former Article 90) of the EC Treaty* (Oxford: Oxford University Press).

Bundeskartellamt (2002) 'Contribution to ICN Subgroup Analytical Framework for Merger Review' 29 April http://www.internationalcompetitionnetwork.org/afsggermany.pdf 1–15.

Burnet, Edward (1984) 'Streamlining Antitrust Litigation by the 'Facial Examination' of Restraints: The Burger Court and the Per Se Rule of Reason Distinction' 60 *Wash L Rev* 1–32.

Buttigieg, Eugene (2005) 'Consumer Interests under the EC's Competition Rules on Collusive Practices' 16 *EBLR* 643–718.

BUTZ, DAVID A, and ANDREW N KLEIT (2001) 'Are Vertical Restraints Pro- or Anticompetitive? Lessons from Interstate Circuit' 44 *J Law Econ* 131–159.

CALDWELL, BRUCE J (1991) 'Clarifying Popper' 29 *J Econ Lit* 1–33.

CALKINS, STEPHEN (2000) 'California Dental Association: Not A Quick Look but Not the Full Monty' 67 *Antitrust LJ* 495–557.

CALVANI, TERRY (2003) 'Devolution and Convergence in Competition Enforcement' 24 *ECLR* 415–423.

CAMESASCA, PETER D (1999) 'The Explicit Efficiency Defence in Merger Control: Does It Make the Difference?' 20 *ECLR* 14–28.

CAMPBELL, DAVID, ed. (2001) *The Relational Theory of Contract: Selected Works of Ian Macneil* (London: Sweet & Maxwell).

CAPLAN, BRYAN, and EDWARD STRINGHAM (2003) 'Networks, Law, and the Paradox of Cooperation' 16 *Review of Austrian Economics* 309–326.

CAPOBIANCO, ANTONIO (2004) 'Information Exchange under EC Competition Law' 41 *CML Rev* 1247–1276.

CAPPELLETTI, MAURO (1987) 'Is the Court of Justice "Running Wild?"' 12 *ELRev* 3–17.

CARCHEDI, GUGLIELMO (2001) *For Another Europe: A Class Analysis of European Economic Integration* (London: Verso).

CARLIN, FIONA M (1996) 'Vertical Restraints: Time for Change?' 17 *ECLR* 283–288.

CARLTON, DENNIS W, and JEFFREY M PERLOFF (2000) *Modern Industrial Organization* (Third Edition Harlow: Addison-Wesley).

CARRIER, MICHAEL A (1999) 'The Real Rule of Reason: Bridging the Disconnect' *BYU L Rev* 1665–1365.

CASE, KARL E, and RAY C FAIR (1999) *Principles of Economics* (Fifth Edition Upper Saddle River, New Jersey: Prentice Hall).

CASS, RONALD A, and KEITH N HYLTON (2001) 'Antitrust Intent' 74 *S Cal L Rev* 657–745.

CASTAÑEDA, GABRIEL (1998) 'Competition Policy Objectives' in Ehlermann, Claus Dieter and Laraine L Laudati (eds) *European Competition Law Annual 1997: The Objectives of Competition Policy* (Oxford: Hart) 41–52.

CHAMBERLIN, EDWARD H (1948) 'Proportionality, Divisibility and Economies of Scale' 62 *Quart J Econ* 229–262.

——(1949) 'Reply: Proportionality, Divisibility, and Economies of Scale: Two Comments' 63 *Quart J Econ* 137–143.

——(1950) 'Product Heterogeneity and Public Policy' 40 *Amer Econ Rev* 85–92.

CHANTRY, LEN (1991) 'Intention and Purpose in Criminal Law' 13 *Liverpool LR* 37–52.

CHUNG, C M (1995) 'The Relationship between State Regulation and EC Competition Law: Two Proposals for a Coherent Approach' 16 *ECLR* 87.

CINI, MICHELLE, and LEE MCGOWAN (1998) *Competition Policy in the European Union* (Basingstoke: Macmillan).

CLARK, JOHN MAURICE (1961) *Competition as a Dynamic Process* (Washington: Brookings Institution).

Clarkson, Kenneth W, and Roger Leroy Miller (1982) *Industrial Organization: Theory, Evidence, and Public Policy* (New York: McGraw-Hill).
Coase, R H (1937) 'The Nature of the Firm' 4 *Economica* 386–405.
——(1994) 'Economics and Contiguous Disciplines' in *Essays on Economics and Economists* (Chicago: University of Chicago Press).
Coate, Malcolm B, and Jeffrey H Fischer (2001) 'Can Post-Chicago Economics Survive Daubert?' 34 *Akron L Rev* 795–852.
Collins, Wayne D (1984) 'Efficiency and Equity in Vertical Competition Law: Balancing the Tensions in the EEC and the United States' in Hawk, Barry E (ed) *Annual Proceedings of the Fordham Corporate Law Institute: International Antitrust Law & Policy 1983* (New York; NY: Transnational Juris) 501–527.
——(1999) 'California Dental Association and the Future of the Rule of Reason Analysis' 14 *Antitrust* 54–62.
Comanor, William S (1967) 'Vertical Mergers, Market Powers, and Antitrust Laws' 57 *Amer Econ Rev* 254–265.
——(1968) 'Vertical Territorial and Customer Restrictions: White Motor and Its Aftermath' 81 *Harv L Rev* 1419–1438.
——(1985) 'Vertical Price-Fixing, Vertical Market Restrictions, and the New Antitrust Policy' 98 *Harv L Rev* 983–1002.
—— and H E Frech III (1985) 'The Competitive Effects of Vertical Agreements?' 75 *Amer Econ Rev* 539–546.
Comments (1952) 'The Per Se Illegality of Price Fixing: Sans Power, Purpose, or Effect' 19 *U Chi L Rev* 837–868.
Constance, Lisa M (1999) 'Antitrust Law—Exercising the Rule of Reason: Supreme Court Revises Analysis of Vertical Maximum Price-Fixing, Bringing It Closer to Achieving the Goals of the Sherman Act—State Oil Co v Khan' 72 *Temp L Rev* 467–492.
Cooke, Judge John D (2000) 'Changing Responsibilities and Relationships for Community and National Courts: The Implications of the White Paper' in 4, CELS Occasional Paper No. (ed) *The Modernisation of European Competition Law: The Next Ten Years* (Cambridge: University Printing Services, University Press) 58–73.
Coppi, Lorenzo, and Paul Dobson (2002) 'The Importance of Market Conduct in the Economic Analysis of Mergers-the Interbrew Saga' 23 *ECLR* 386–391.
Corden, W M (1972) 'Economies of Scale and Customs Union Theory' 80 *J Polit Economy* 465–475.
Cornish, W R (1996) *Intellectual Property* (Third Edition London: Sweet & Maxwell).
Cosma, Hakon A, and Richard Whish (2003) 'Soft Law in the Field of EU Competition Policy' 14 *EBLR* 25–56.
Cowen, Tyler, and Daniel Sutter (1999) 'The Costs of Cooperation' 12 *Review of Austrian Economics* 161–173.
————(2005) 'Conflict, Cooperation and Competition in Anarchy' 18 *Review of Austrian Economics* 109–115.
Craig, P P (1991) 'Constitutions, Property and Regulation' *Public Law* 538–554.

CRAIG, PAUL (1997) 'Public Law and Control over Private Power' in Taggart, Michael (ed) *The Province of Administrative Law* (Oxford: Hart) 196–216.

—— and GRÀINNE de BÚRCA (1998) *EU Law: Text, Cases, and Materials* (Second Edition Oxford: Oxford University Press).

——(2001) 'Constitutions, Constitutionalism, and the European Union' 7 *ELJ* 125–150.

—— and GRÀINNE de BÚRCA (2003) *EU Law: Text, Cases, and Materials* (Third Edition Oxford: Oxford University Press).

CRUZ, JULIO BAQUERO (2002) *Between Competition and Free Movement: The Economic Constitutional Law of the European Community* (Oxford: Hart).

CUNNINGHAM, JAMES PATRICK (1973) *The Competition Law of the EEC: A Practical Guide* (London: K Page).

DALTON, JAMES A, and STANFORD L LEVIN (1973) 'Allocative and Distributive Effects of Monopoly' in Dalton, James A and Stanford L Levin (eds) *The Antitrust Dilemma* (Lexington, Massachusetts: Lexington Books).

DALY, HERMAN E, and JOHN B COBB (1990) *For the Common Good: Redirecting the Economy toward Community, the Environment, and a Sustainable Future* (London: Green Print).

DASHWOOD, ALAN, DIANA GUY, HUGH LADDIE, and GUY I F LEIGH (1976) 'The Sugar Industry Marathon' 3 *ELRev* 479–499.

——(1996) 'The Limits of European Community Powers' 21 *ELRev* 113–128.

——(1998) 'States in the European Union' 23 *ELRev* 201–216.

——(2004) 'The Relationship between the Member States and the European Union/European Community' 41 *CML Rev* 355–381.

DAVIES, GARETH (2002) 'Welfare as a Service' 29 *LIEI* 27–40.

DEARDS, ELSPETH (2002) 'Closed Shop Versus One Stop Shop: The Battle Goes On' 27 *ELRev* 618–627.

DECKERT, MARTINA (2000) 'Some Preliminary Remarks on the Limitations of European Competition Law' 8 *ERPL* 173–185.

DEMSETZ, HAROLD (1973) *The Market Concentration Doctrine: An Examination of Evidence and a Discussion of Policy* (Washington: American Enterprise Institute for Public Policy Research).

——(1982) 'Barriers to Entry' 72 *Amer Econ Rev* 47–57.

DENMAN, DANIEL (2002) 'Case Comment: European Court of Justice: Anti-Competitive Agreements—Professional Rules—Netherlands Bar' 23 *ECLR* N73.

DENNING, LORD (1961) *Responsibility before the Law* (Jerusalem: Magnes Press).

DEPRANO, MICHAEL E, and JEFFREY B NUDENT (1969) 'Economies as an Antitrust Defense' 59 *Amer Econ Rev* 947–953.

DERINGER, ARVED (1962) *Some Practical Aspects of the Antitrust Provisions of the Treaty of Rome* (London: British Institute of International and Comparative Law).

——(1968) *The Competition Law of the European Economic Community: A Commentary on the EEC Rules of Competition (Articles 85 to 90) Including the Implementing Regulations and Directives* (New York: Commerce Clearing House).

Dibadj, Reza (2004) 'Saving Antitrust' 75 *University of Colorado Law Review* 745–862.
Diwan, Romesh K (1966) 'Alternative Specifications of Economies of Scale' 33 *Economica* 442–453.
Dixit, A K (1980) 'The Role of Investment in Entry-Deterrence' 90 *Econ J* 95–106.
Dobson, Paul W, and Michael Waterson (1996) *Vertical Restraints and Competition Policy* (London: Office of Fair Trading).
——— (1997) 'Countervailing Power and Consumer Prices' 107 *Economic Journal* 418–430.
Dougan, Michael (2000) 'Minimum Harmonization and the Internal Market' 37 *CML Rev* 853–885.
Douglas-Scott, Sionaidh (2002) *Constitutional Law of the European Union* (Harlow: Pearson Education).
Drijber, Berend Jan (2005) 'Joined Cases C-264/01, C-306/01, C-354/01 and C-355/01, AOK Bundesverband A.O, Judgment of the Full Court 16 March 2004, Not yet Reported' 42 *CML Rev* 523–533.
Duhamel, Marc, and Peter G C Townley (2003) 'An Effective and Enforceable Alternative to the Consumer Surplus Standard' 26 *W Comp* 3–24.
Duns, John (1994) 'Competition Law and Public Benefits' 16 *Adelaide LR* 245–267.
Dutton, Harold Irvin (1984) *The Patent System and Inventive Activity During the Industrial Revolution 1750–1852* (Manchester: Manchester University Press).
Easterbrook, Frank H (1984) 'The Limits of Antitrust' 63 *Tex L Rev* 1–40.
—— (1987) 'Allocating Antitrust Decision Making Tasks' 76 *Geo LJ* 305–320.
Economist (1999) 'Antitrust on Trial' November 13 *Economist* 122.
—— (2000a) 'Bill Rockefeller?' April 29 *Economist* 16.
—— (2000b) 'The New Enforcers' October 7 *Economist* 125–130.
—— (2000c) 'The Knowledge Monopolies' April 8 *Economist* 95–99.
—— (2000d) 'Trust and Antitrust' October 7 *Economist* 21.
—— (2000e) 'Spaghetti Monti' October 7 *Economist* 129.
—— (2000f) 'Barriers Real and Imagined' December 9 *Economist* 132.
—— (2001) 'Copyrights and Wrongs' April 28 *Economist* 82–83.
—— (2002) 'The Growth Machine' May 18 *Economist* 74.
—— (2004) 'Torturing Your Rivals' August 28 *Economist* 58.
Ehlermann, Claus-Dieter (1996) 'Implementation of EC Competition Law by National Anti-Trust Authorities' 17 *ECLR* 88–95.
—— (1995) 'Reflections on a European Cartel Office' 32 *CML Rev* 471–486.
—— and Berend Jan Drijber (1996) 'Legal Protection of Enterprises: Administrative Procedure, in Particular Access to Files and Confidentiality' 17 *ECLR* 375–383.
—— (2000) 'The Modernisation of EC Antitrust Policy: A Legal and Cultural Revolution' 37 *CML Rev* 537–590.
—— and Isabela Atanasiu (2002) 'The Modernisation of EC Antitrust Law: Consequences for the Future Role and Function of the EC Courts' 23 *ECLR* 72–80.

EHLERMANN, CLAUS-DIETER, and LARAINE L LAUDATI, eds. (1998) *European Competition Law Annual 1997: The Objectives of Competition Policy* (Oxford: Hart Publishing).

EISNER, MARC ALLEN (1991) *Antitrust and the Triumph of Economics: Institutions, Expertise, and Policy Change* (Chapel Hill: University of North Carolina Press).

ELHAUGE, EINER (2002) 'Soft on Microsoft: The Potemkin Antitrust Settlement' March 25 *The Weekly Standard.*

ELLIS, JOSEPH J A (1963) 'Source Material for Article 85(1) EEC' *Fordham Law Review* 247–278.

ELZINGA, KENNETH G (1977) 'The Goals of Antitrust: Other Than Competition and Efficiency, What Else Counts?' 125 *U Pa L Rev* 1191–1213.

——(1989) 'Unmasking Monopoly: Four Types of Economic Evidence' in Larner, Robert J and James W Meehan Jr. (eds) *Economics and Antitrust Policy* (New York: Quorum).

ENCAOUA, DAVID, PAUL GEROSKI, and ALEXIS JACQUEMIN (1986) 'Strategic Competition and the Persistence of Dominant Firms: A Survey' in Stiglitz, Joseph E and G Frank Mathewson (eds) *New Developments in the Analysis of Market Structure: Proceedings of a Conference Held by the International Economic Association in Ottawa, Canada* (London: Macmillan).

ESTRIN, SAUL, ed. (1994) *Privatization in Central and Eastern Europe* (London: Longman).

EUCKEN, WALTER (1989) 'What Kind of Economic and Social System' in Peacock, Alan and Hans Willgerodt (eds) *Germany's Social Market Economy: Origins and Evolution* (Basingstoke: Macmillan).

EVANS, DAVID S (2001) 'Dodging the Consumer Harm Inquiry: A Brief Survey of Recent Government Antitrust Cases' 75 *St John's L Rev* 545–559.

EVERITT, NICHOLAS, and ALEC FISHER (1995) *Modern Epistemology: A New Introduction* (London: McGraw-Hill).

FARRELL, JOSEPH, and MATTHEW RABIN (1996) 'Cheap Talk' 10 *Journal of Economic Perspectives* 103–118.

FARRELL, M J (1957) 'The Measurement of Productive Efficiency' 120 *Journal Of The Royal Statistical Society Series A (General)* 253–290.

FAULL, JONATHAN, and ALI NIKPAY, eds. (1999) *Faull & Nikpay: The EC Law of Competition* (Oxford: Oxford University Press).

FELDMAN, PAUL (1971) 'Efficiency, Distribution, and the Role of Government in A Market Economy' 79 *J Polit Economy* 508–526.

FELS, ALLAN, and GEOFF EDWARDS (1998) 'Competition Policy Objectives' in Ehlermann, Claus Dieter and Laraine L Laudati (eds) *European Competition Law Annual 1997: The Objectives of Competition Policy* (Oxford: Hart) 53–65.

FERRY, J E (1981) 'Selective Distribution and Other Post-Sales Restrictions' 2 *ECLR* 209–217.

FISHER, FRANKLIN M (1985) 'The Social Costs of Monopoly and Regulation: Posner Reconsidered' 93 *J Polit Economy* 410–416.

FITZSIMMONS, ANTHONY (1994) *Insurance Competition Law: A Handbook to the Competition Law of the European Union and the European Economic Area* (London: Graham and Trotman).

FORRESTER, IAN, and CHRISTOPHER NORALL (1984) 'The Laicization of Community Law: Self Help and the Rule of Reason: How Competition Law Is and Could Be Applied' 21 *CML Rev* 11–51.

FORRESTER QC, IAN S (1998) 'Competition Law Implementation at Present' in Ehlermann, Claus Dieter and Laraine L Laudati (eds) *European Competition Law Annual 1997: The Objectives of Competition Policy* (Oxford: Hart) 359–385.

——(2001) 'The Reform of the Implementation of Articles 81 and 82 Following Publication of the Draft Regulation' 28 *LIEI* 173–194.

—— and JACQUELYN F MACLENNAN (2003) 'EC Competition Law 2001–2002' 22 *YEL* 499–581.

FOX, ELEANOR M, and LAWRENCE A SULLIVAN (1987) 'Antitrust—Retrospective and Perspective: Where Are We Coming From? Where Are We Going?' 62 *NYU L Rev* 936–988.

——(2001) 'The Elusive Promise of Modernisation: Europe and the World' 28 *LIEI* 141–149.

FRANTZ, ROGER (1992) 'X-Efficiency and Allocative Efficiency: What Have We Learned?' 82 *Amer Econ Rev* 434–438.

FRANTZ, ROGER S (1988) *X-Efficiency: Theory, Evidence and Applications* (Boston: Kluwer).

FREEDLAND, MARK (1994) 'Government by Contract and Private Law' *Public Law* 86–104.

FRIED, CHARLES (1981) *Contract as Promise: A Theory of Contractual Obligation* (Cambridge, Mass.: Harvard University Press).

FRIEDMAN, MILTON, and ROSE FRIEDMAN (1980) *Free to Choose: A Personal Statement* (London: Harcourt).

FUDENBERG, D, and J TIROLE (1986) 'A Signal Jamming Theory of Predation' 17 *RAND J Econ* 366–376.

FULLER, LON L, and WILLIAM R PERDUE Jr (1936a) 'The Reliance Interest in Contract Damages:1' 46 *Yale LJ* 52–96.

———— (1936b) 'The Reliance Interest in Contract Damages:2' 46 *Yale LJ* 373–420.

——(1965) 'A Reply to Professors Cohen and Dworkin' 10 *Villanova Law Review* 655–666.

——(1969) *The Morality of Law* (Revised Edition New Haven: Yale University Press).

FURSE, MARK (1996) 'The Role of Competition Policy: A Survey' 17 *ECLR* 250–257.

——(2000) *Competition Law of the UK & EC* (Second Edition London: Blackstone).

——(2004) *Competition Law of the UK & EC* (Fourth Edition London: Blackstone).

GAAY FORTMAN, BASTIAAN DE (1966) *Theory of Competition Policy: A Confrontation of Economic, Political, and Legal Principles* (Amsterdam: North-Holland).

GAGLIARDI, ANDREA FILIPPO (2000) 'United States and European Union Antitrust Versus State Regulation of the Economy: Is There a Better Test?' 25 *ELRev* 353–373.

GALBRAITH, JOHN KENNETH (1954) 'Countervailing Power' 44 *Amer Econ Rev* 1–6.

——(1957) *American Capitalism: The Concept of Countervailing Power* (Revised Edition London: Hamish Hamilton).

——(1959) 'Mr Hunter on Countervailing Power: A Comment' 69 *Econ J* 168–170.

——(2004) *The Economics of Innocent Fraud: Truth for Our Time* (London: Penguin).

GAVIL, ANDREW I (2000a) 'Copperweld 2000: The Vanishing Gap between Sections 1 and 2 of the Sherman Act' 68 *Antitrust LJ* 87–110.

——(2000b) 'Editor's Note' 68 *Antitrust LJ* 331–335.

—— WILLIAM E KOVACIC, and JONATHAN B BAKER (2002) *Antitrust Law in Perspective: Cases, Concepts, and Problems in Competition Policy* (St Paul, MN: West).

GELLHORN, ERNEST, and TERESA TATHAM (1984/1985) 'Making Sense out of the Rule of Reason' 35 *Case W Res L Rev* 155–182.

—— WILLIAM E KOVACIC, and STEPHEN CALKINS (2004) *Antitrust Law and Economics in a Nutshell* (Fifth Edition St. Paul, MN: West).

GERBER, DAVID J (1994) 'The Transformation of European Community Competition Law?' 35 *Harvard International Law Journal* 97–147.

——(2001) *Law and Competition in Twentieth Century Europe: Protecting Prometheus* (Paperback edition Oxford: Oxford University Press).

GERHART, PETER M (1981) 'The "Competitive Advantages" Explanation for Intrabrand Restraints: An Antitrust Analysis' *Duke LJ* 417–448.

GEROSKI, PAUL, and JOACHIM SCHWALBACH (1989) *Barriers to Entry and Intensity of Competition in European Markets* (Luxembourg: Office for Official Publications of the European Communities).

GETTIER, EDMUND L (1967) 'Is Justified True Belief Knowledge' in Griffiths, A Phillips (ed) *Knowledge and Belief* (London: Oxford University Press) 144–146.

GIFFORD, DANIEL J (1986) 'The Role of the Ninth Circuit in the Development of the Law of Attempt to Monopolize' 61 *Notre Dame L Rev* 1021–1051.

GIJLSTRA, D J, and D F MURPHY (1977) 'Some Observations on the Sugar Cases' 14 *CML Rev* 45–71.

GIVENS, RICHARD A (1983) *Antitrust: An Economic Approach* (New York, New York: Law Journal Seminars-Press).

GLEISS, ALFRED, and MARTIN HIRSCH (1981) *Common Market Cartel Law* (Third Edition Washington DC: Bureau of National Affairs).

GOLDMAN QC, CALVIN S, and MILOS BARUTCISKI (1998) 'Competition Law Implementation at Present' in Ehlermann, Claus Dieter and Laraine L Laudati (eds) *European Competition Law Annual 1997: The Objectives of Competition Policy* (Oxford: Hart) 387–416.

GONZALEZ-DIAZ, ENRIQUE (1996) 'The Notion of Ancillary Restraints under EC Competition Law' 19 *Fordham Int'l LJ* 951–998.

GORDON, GERALD, and EDWARD V MORSE (1975) 'Evaluation Research' 1 *Annual Review of Sociology* 339–361.

GORMLEY, LAURENCE W (2002) 'Competition and Free Movement: Is the Internal Market the Same as the Common Market?' 13 *EBLR* 517–522.

GOULD, J R, and B S YAMEY (1967) 'Professor Bork on Vertical Price Fixing' 76 *Yale LJ* 722–730.

GOYDER, D G (1998) *EC Competition Law* (Third Edition Oxford: Oxford University Press).

——(2003) *EC Competition Law* (Fourth Edition Oxford: Oxford University Press).

GOYDER, DAN (1993) 'User-Friendly Competition Law' in Slot, Piet Jan and Alison McDonnell (eds) *Procedure and Enforcement in EC and US Competition Law* (London: Sweet & Maxwell) 1–5.

GREEN, NICHOLAS (1988) 'Article 85 in Perspective: Stretching Jurisdiction, Narrowing the Concept of a Restriction and Plugging a Few Gaps' 9 *ELRev* 190–206.

GREGORY, JULIAN (2005) 'From Rio to Meca: Another Step on the Winding Road of Competition Law and Sport' 64 *Cambridge Law Journal* 51–54.

GREWLICH, ALEXANDRE S (2001) 'Globalisation and Conflict in Competition Law: Elements of Possible Solutions' 24 *W Comp* 367–404.

GRIFFITHS, MARK (2000) 'A Glorification of de Minimis—the Regulation on Vertical Agreements' 21 *ECLR* 241–247.

GRIMES, WARREN S (1995) 'Brand Marketing, Intrabrand Competition and the Multibrand Retailer: The Antitrust Law of Vertical Restraints' 64 *Antitrust LJ* 83–136.

——(1998) 'Making Sense of State Oil Co v Khan: Vertical Maximum Price Fixing under A Rule of Reason' 66 *Antitrust LJ* 567–611.

GROSSKOPF, S (1993) 'Efficiency and Productivity' in Fried, Harold O, C A Knox Lovell and Shelton S Schmidt (eds) *The Measurement of Productive Efficiency* (Oxford: Oxford University Press).

GROVES, THEODORE, and JOHN LEDYARD (1977) 'Optimal Allocation of Public Goods: A Solution to The "Free Rider" Problem' 45 *Econometrica* 783–810.

GUERRIN, MAURICE, and GEORGIOS KYRIAZIS (1993) 'Cartels: Proof and Procedural Issues' in Hawk, Barry E (ed) *Annual Proceedings of the Fordham Corporate Law Institute: International Antitrust Law & Policy 1992* (New York; NY: Transnational Juris) 773–843.

GYSELEN, LUC (1994) 'Anti-Competitive State Measures under the EC Treaty: Towards a Substantial Legality Standard' CC *ELRev* 55–89.

——(2000) 'Case C-67/96, Albany v Stichting Bedrijfspensioenfonds Textielindustrie; Joined Cases C-115–117/97, Brentjens' Handelsonderneming v Stichting Bedrijfspensioenfonds Voor de Handel in Bouwmaterialen; and Case C-219/97, Drijvende Bokken v Stichting Pensioenfonds Voor de Vervoeren Havenbedrijven' 37 CML Rev 425–448.

——(2002) 'The Substantive Legality Test under Article 81(3) EC Treaty—Revisited in Light of the Commission's Modernization Initiative' in Bogdandy,

Armin von, Petros C Mavroidis and Yves Mény (eds) *European Integration and International Co-Ordination: Studies in Transnational Economic Law in Honour of Claus-Dieter Ehlermann* (The Hague: Kluwer Law International) 181–197.

Haffner, Alex (2005) 'United States Postal Service v Flamingo Industries (USA) Ltd: Does the USA Have the Answer to The "Undertaking Problem" In Europe' 26 *ECLR* 397–402.

Hahn, Verena (2000) 'Antitrust Enforcement: Abuse Control or Notification' 10 *Europ J Law Econ* 69–91.

Hamilton, Clive (2004) *Growth Fetish* (London: Pluto).

Hancher, Leigh, and Jose-Luis Buendia Sierra (1998) 'Cross-Subsidization and EC Law' 35 *CML Rev* 901–945.

Harberger, Arnold C (1971) 'Three Basic Postulates for Applied Welfare Economics: An Interpretive Essay' 9 *J Econ Lit* 785–797.

——(1978) 'On the Use of Distributional Weights in Social Cost-Benefit Analysis' 86 *J Polit Economy* S87-S120.

——(1980) 'Reply to Layard and Squire' 88 *J Polit Economy* 1050–1052.

Harding, Christopher, and Julian Joshua (2003) *Regulating Cartels in Europe: A Study of Legal Control of Corporate Delinquency* (Oxford: Oxford University Press).

Harlow, Carol, and Richard Rawlings (1997) *Law and Administration* (Second Edition London: Butterworths).

Harrison, Jeffrey L (2004) 'Rationalizing the Allocative/Distributive Relationship in Copyright' 32 *Hofstra Law Review* 853–906.

Hart, H L A, and Honoré, A M (1959) *Causation in the Law* (Oxford: Clarendon Press).

Hartley, Trevor C (1996) 'The European Court, Judicial Objectivity and the Constitution of the European Union' 112 LQR 95–109.

——(1998) *The Foundations of European Community Law: An Introduction to the Constitutional and Administrative Law of the European Community* (Fourth Edition Oxford: Clarendon Press).

Hatzopoulos, Vassilis (2000) 'Recent Developments of the Case Law of the ECJ in the Field of Services' 37 *CML Rev* 43–82.

——(2002) 'Killing National Health and Insurance Systems but Healing Patients? The European Market for Health Care Services after the Judgments of the ECJ in Vanbraekel and Peerbooms' 39 *CML Rev* 683–729.

Hausman, Daniel M, and Michael S McPherson (1996) *Economic Analysis and Moral Philosophy* (Cambridge: Cambridge University Press).

Hawk, Barry E (1985) *United States, Common Market, and International Antitrust: A Comparative Guide* (Second Edition Clifton, New Jersey: Prentice Hall Law & Business Inc/Harcourt Brace Jovanovich).

——(1988) 'The American (Anti-Trust) Revolution: Lessons for the EEC?' 9 ECLR 53–87.

——(1995) 'System Failure: Vertical Restraints and EC Competition Law' 32 *CML Rev* 973–989.

HAY, DONALD A (1999) 'Is More Like Europe Better?: An Economic Evaluation of Recent Changes in UK Competition Policy' in Green, Nicholas and Aidan Robertson (eds) *The Europeanisation of UK Competition Law* (Oxford: Hart) 35–59.

HAY, GEORGE A (1991) 'Market Power in Antitrust' 60 *Antitrust* LJ 807–827.

——(2000) 'The Meaning Of "Agreement" Under the Sherman Act: Thoughts from The "Facilitating Practices" Experience' 16 *Rev Ind Organ* 113–129.

HAYEK, FRIEDRICH A VON (1980) *Individualism and Economic Order* (Chicago: University of Chicago Press).

HAYS, THOMAS (2001) 'Anti-Competitive Agreements and Extra-Market Parallel Importation' 26 *ELRev* 468–488.

HELLEINER, KARL F (1951) 'Moral Conditions of Economic Growth' 11 *Journal of Economic History* 97–116.

HERTZ, NOREENA (2002) *The Silent Takeover: Global Capitalism and the Death of Democracy* (London: Arrow).

HERVEY, TAMARA (2000) 'Social Solidarity: A Buttress against Internal Market Law?' in Shaw, Jo (ed) *Social Law and Policy in an Evolving European Union* (Oxford: Hart).

HEYDON, J D (1999) *The Restraint of Trade Doctrine* (Second Edition Sydney: Butterworths).

HICKS, J R (1935) 'Annual Survey of Economic Theory: The Theory of Monopoly' 3 *Econometrica* 1–20.

HILDEBRAND, DORIS (1998) *The Role of Economic Analysis in the EC Competition Rules* (The Hague: Kluwer Law International).

——(2002a) 'The European School in EC Competition Law' 25 *W Comp* 3–23.

——(2002b) *The Role of Economic Analysis in the EC Competition Rules: The European School* (Second Edition The Hague: Kluwer Law International).

HOLMES, KATHERINE (2000) 'The EC White Paper on Modernisation' 23 *W Comp* 51–79.

HOMAN, PAUL T (1956) 'The Social Goals of Economic Growth in the United States' 46 *Amer Econ Rev* 24–34.

HONIG, FREDERICK, WILLIAM J BROWN, ALFRED GLEISS, and MARTIN HIRSCH (1963) *Cartel Law of the European Economic Community* (London: Butterworths).

HONORÉ, TONY (1999) *Responsibility and Fault* (Oxford: Hart).

HORNSBY, STEPHEN (1987) 'Competition Policy in the 80's: More Policy Less Competition' 12 *ELRev* 79–101.

House Of Lords Select Committee (2000) Fourth Report: Reforming EC Competition Procedures HL 33.

HOUTTE, BEN VAN (1982) 'A Standard of Reason in EEC Antitrust Law: Some Comments on the Application of Parts 1 and 3 of Article 85' 4 *Journal Of International Law* & Business 497–516.

HUGHES, EDWIN J (1994) 'The Left Side of Antitrust: What Fairness Means and Why It Matters' 77 *Marq L Rev* 265–306.

HUMMEL, JEFFRY ROGERS (1990) 'National Goods Versus Public Goods: Defense, Disarmament, and Free Riders' 4 *The Review Of Austrian Economics* 88–122.

HUNT, MURRAY (1997) 'Constitutionalism and the Contractualisation of Government' in Taggart, Michael (ed) *The Province of Administrative Law* (Oxford: Hart) 21–39.

HUNTER, ALEX (1958) 'Notes on Countervailing Power' 68 *Econ J* 89–103.

——(1959) 'Galbraith on Countervailing Power: A Reply' 69 *Econ J* 822–823.

HURLEY, EILEEN R (1987) 'Pronuptia de Paris v Schillgalis: Permissible Restraints of Trade on Franchising in the EEC' 8 *Journal Of International Law & Business* 476–502.

HYLTON, KEITH N (2003) *Antitrust Law: Economic Theory and Common Law Evolution* (Cambridge: Cambridge University Press).

IBBETSON, DAVID J (1999) *A Historical Introduction to the Law of Obligations* (Oxford: Oxford University Press).

ICHINO, PETER (2001) 'Collective Bargaining and Antitrust Laws: An Open Issue' 17 *International Journal of Comparative Labour Law and Industrial Relations* 185–198.

ICN Merger Analytical Framework Subgroup (2004) 'Efficiencies' April http://www.internationalcompetitionnetwork.org/seoul/amg_chap6_efficiencies.pdf 1–38.

JACOBS, MICHAEL S (1995) 'An Essay on the Normative Foundations of Antitrust Economics' 74 *NC L Rev* 219–266.

JAKOBSEN, PETER STIG, and MORTEN BROBERG (2002) 'The Concept of Agreement in Article 81 EC: On the Manufacturer's Right to Prevent Parallel Trade within the European Community' 23 *ECLR* 127–141.

JEBSEN, PER, and ROBERT STEVENS (1996) 'Assumptions, Goals and Dominant Undertakings: The Regulation of Competition under Article 86 of the European Union' 64 *Antitrust LJ* 445–516.

JENNY, FRÉDÉRIC (1994) 'Competition and Efficiency' in Hawk, Barry E (ed) *Annual Proceedings of the Fordham Corporate Law Institute: International Antitrust Law & Policy 1993* (New York; NY: Transnational Juris) 185–220.

——(1999) 'Competition Law Enforcement: Is Economic Expertise Necessary' in Hawk, Barry E (ed) *Annual Proceedings of the Fordham Corporate Law Institute: International Antitrust Law & Policy 1998* (Yonkers, New York: Juris Publications) 185–201.

JEPHCOTT, MARK, and THOMAS LÜBBIG (2003) *Law of Cartels* (Bristol: Jordans).

JOFFE, ROBERT D (2001) 'Antitrust Law and Proof of Consumer Injury' 75 *St John's L Rev* 615–632.

JOLIET, RENÉ (1967) *The Rule of Reason in Antitrust Law: American, German and Common Market Laws in Comparative Perspective* (Liège: Faculté de droit).

JONES, ALISON (1993) 'Woodpulp: Concerted Practice and/or Conscious Parallelism?' 14 *ECLR* 273.

—— and BRENDA SUFRIN (2001) *EC Competition Law: Text, Cases, and Materials* (Oxford: Oxford University Press).

—— and BRENDA SUFRIN (2004) *EC Competition Law: Text, Cases, and Materials* (Second Edition Oxford: Oxford University Press).

JONES, CHARLES I (1998) *Introduction to Economic Growth* (London: W W Norton).
JOSHUA, JULIAN MATHIC (1987) 'Proof in Contested EEC Competition Cases: A Comparison with the Rules of Evidence in Common Law' 12 *EL Rev* 315–353.
—— and SARAH JORDAN (2004) 'Combinations, Concerted Practices and Cartels: Adopting the Concept of Conspiracy in European Community Competition Law' 24 *Nw J Int'l L & Bus* 647–681.
KAHN, ALFRED E (1953) 'Standards for Antitrust Policy' 67 *Harv L Rev* 28–54.
KALDOR, NICHOLAS (1939) 'Welfare Propositions of Economics and Interpersonal Comparisons of Utility' 49 *Econ J* 549–552.
KAPLAN, DAVID P (1998) 'The Nuts and Bolts of Antitrust Analysis: Some Thoughts on How to Develop the Facts' in McChesney, Fred S (ed) *Economic Inputs, Legal Outputs: The Role of Economists in Modern Antitrust* (Chichester: Wiley).
KAPLOW, LOUIS, and STEVEN SHAVELL (2002) *Fairness Versus Welfare* (London: Harvard University Press).
KATTAN, JOSEPH (1996) 'The Role of Efficiency Considerations in the Federal Trade Commission's Antitrust Analysis' 64 *Antitrust LJ* 613–632.
KAUPER, THOMAS E (1993) 'The Use of Consent Decrees in American Antitrust Cases' in Slot, Piet Jan and Alison McDonnell (eds) *Procedure and Enforcement in EC and US Competition Law* (London: Sweet & Maxwell) 104–113.
—— (1997) 'The Problem of Market Definition under EC Competition Law' 20 *Fordham Int'l LJ* 1682–1767.
KAYSEN, CARL (1951) 'Collusion under the Sherman Act 1' 65 *Quart J Econ* 263–270.
—— and DONALD F TURNER (1959) *Antitrust Policy: An Economic and Legal Analysis* (Cambridge, Mass: Harvard University Press).
KELLAWAY, ROSALIND (1997) 'Vertical Restraints: Which Option?' 18 *ECLR* 387–391.
KELLY, KENNETH (1988) 'The Role of the Free Rider in Resale Price Maintenance: The Loch Ness Monster of Antitrust Captured' 10 *Geo Mason L Rev* 327–381.
KENDRICK, JOHN W (1977) *Understanding Productivity: An Introduction to the Dynamics of Productivity Change* (Baltimore: Johns Hopkins University Press).
KENNEDY, DUNCAN (1982) 'The Stages of Decline of the Public/Private Distinction' 130 *U Pa L Rev* 1349–1357.
KENNY, ANTHONY (1966) 'Intention and Purpose' 63 *The Journal Of Philosophy* 642–651.
KERSE, CHRISTOPHER S (1998) EC *Antitrust Procedure* (Fourth Edition London: Sweet & Maxwell).
—— and NICHOLAS KHAN (2005) EC *Antitrust Procedure* (Fifth Edition London: Sweet & Maxwell).
KEYTE, JAMES A (1997) 'What It Is and How It Is Being Applied: The "Quick Look" Rule of Reason' 11—summer *Antitrust* 21–25.
—— and NEAL R STOLL (2004) 'Markets? We Don't Need No Stinking Markets! The FTC and Market Definition' 49 *Antitrust Bulletin* 593–632.

KHEMANI, R SHYAM (1998) 'Competition Policy Objectices in the Context of A Multilateral Competition Code' in Ehlermann, Claus Dieter and Laraine L Laudati (eds) *European Competition Law Annual 1997: The Objectives of Competition Policy* (Oxford: Hart) 187–245.

KIMEL, DORI (2003) *From Promise to Contract: Towards a Liberal Theory of Contract* (Oxford: Hart).

KINGSTON, SUZANNE (2001) 'A "New Division of Responsibilities" in the Proposed Regulation to Modernise the Rules Implementing Articles 81 and 82 EC? A Warning Call' 22 *ECLR* 340–350.

KIRCHNER, CHRISTIAN (1998) 'Future Competition Law' in Ehlermann, Claus Dieter and Laraine L Laudati (eds) *European Competition Law Annual 1997: The Objectives of Competition Policy* (Oxford: Hart) 513–523.

KLEIN, BENJAMIN, and KEVIN M MURPHY (1988) 'Vertical Restraints as Contract Enforcement Mechanisms' 31 *J Law Econ* 265–297.

——(1993) 'Market Power in Antitrust: Economic Analysis after Kodak' 3 *Supreme Court Economic Review* 43–92.

KLEIN, NAOMI (2001) *No Logo: No Space, No Choice, No Jobs* (London: Flamingo).

KNIGHT, FRANK HYNEMAN (1935) *The Ethics of Competition, and Other Essays* (London: George Allen & Unwin).

——(1951) *The Economic Organization; with an Article, Notes on Cost and Utility* (New York: Augustus M Kelley).

KOBAYASHI, HIDEAKI (1998) 'Competition Policy Objectives' in Ehlermann, Claus Dieter and Laraine L Laudati (eds) *European Competition Law Annual 1997: The Objectives of Competition Policy* (Oxford: Hart) 81–84.

KOLASKY, WILLIAM J (2002) 'Six Guiding Principles for Antitrust Agencies' March 18 *Presentation to International Bar Association Conference on Competition Law and Policy in a Global Context.*

KOMNINOS, ASSIMAKIS P (2001) 'Arbitration and the Modernisation of European Competition Law Enforcement' 24 *W Comp* 211–238.

——(2002) 'New Prospects for Private Enforcement of EC Competition Law: Courage v Crehan and the Community Right to Damages' 39 *CML Rev* 447–487.

KOOPMANS, T (1986) 'The Role of Law in the Next Stage of European Integration' 35 *Int'l & Comp LQ* 925–931.

KORAH, VALENTINE (1973) 'Concerted Practices: The EEC Dyestuffs Case' 36 *Modern Law Review* 220–226.

——(1984) *Exclusive Dealing Agreements in the EEC: Regulation 67/67 Replaced* (London: European Law Centre).

——(1985) *Patent Licensing and EEC Competition Rules: Regulation 2349/84* (Oxford: ESC Publishing Limited).

——(1986) 'EEC Competition Policy—Legal Form or Economic Efficiency' 39 *CLP* 85–109.

—— and WARWICK ROTHNIE (1992) *Exclusive Distribution and the EEC Competition Rules: Regulations 1983/83 & 1984/83* (Second Edition London: Sweet & Maxwell).

KORAH, VALENTINE (1993a) 'The Judgment in Delimitis—A Milestone Towards A Realistic Assessment of the Effects of an Agreement—or A Damp Squib' 8 *Tul Eur & Civ LF* 17–51.

——(1993b) 'Commentary' 20 *Brook J Int'l L* 161–168.

——(1993c) 'EEC Licensing of Intellectual Property' *Fordham Intell Prop Media & Ent LJ* 55–80.

——(1994) 'Exclusive Purchasing Obligations: Mars v Langnese and Schoeller' 15 *ECLR* 171–175.

——(1997) *An Introductory Guide to EC Competition Law and Practice* (Sixth Edition Oxford: Hart).

——(2000) *An Introductory Guide to EC Competition Law and Practice* (Seventh Edition Oxford: Hart).

——(2002) 'The Interface between Intellectual Property and Antitrust: The European Experience' 69 *Antitrust LJ* 801–839.

—— and DENIS O'SULLIVAN (2002) *Distribution Agreements under the EC Competition Rules* (Oxford: Hart).

——(2004) *An Introductory Guide to EC Competition Law and Practice* (Eighth Edition Oxford: Hart).

KOVACIC, WILLIAM E (1990) 'The Antitrust Paradox Revisited: Robert Bork and the Transformation of Modern Antitrust Policy' 36 *Wayne Law Review* 1413–1471.

——(1993) 'The Identification and Proof of Horizontal Agreements under the Antitrust Laws' 38 *Antitrust Bull* 5–81.

KUHN, KAI-UWE (2001) 'Fighting Collusion by Regulating Communication between Firms' *Economic Policy* 167–204.

LANDE, ROBERT H (1989) 'Chicago's False Foundation: Wealth Transfers (Not Just Efficiency) Should Guide Antitrust' 58 *Antitrust LJ* 631–644.

——(2001) 'Consumer Choice as the Ultimate Goal of Antitrust' 62 *University of Pittsburgh Law Review* 503.

LANDES, WILLIAM M, and RICHARD A POSNER (1981) 'Market Power in Antitrust Cases' 94 *Harv L Rev* 937–996.

——(number 45) 'The Art of Law and Economics: An Autobiographical Essay' *Chicago Law And Economics Working Paper* 1–22.

LASOK, K P E (1982) 'Nungesser v Commission' 3 *ECLR* 133–138.

——(2004) 'When Is an Undertaking Not an Undertaking' 25 *ECLR* 383–385.

LAUFER, WILLIAM S (1992) 'Culpability and the Sentencing of Corporations' 71 *Neb L Rev* 1049–1091.

——(1994) 'Corporate Bodies and Guilty Minds' 43 *Emory LJ* 648–730.

LAYARD, RICHARD (1980) 'On the Use of Distributional Weights in Social Cost-Benefit Analysis' 88 *J Polit Economy* 1041–1047.

LEIBENSTEIN, HARVEY (1966) 'Allocative Efficiency Vs "X-Efficiency"' 56 *Amer Econ Rev* 392–415.

——(1988) 'Foreword' in Frantz, Roger S (ed) *X-Efficiency: Theory, Evidence and Applications* (Boston: Kluwer).

LENAERTS, KOEN, and DAMIEN GERARD (2004) 'Decentralisation of EC Competition Law Enforcement: Judges in the Frontline' 27 *W Comp* 313–349.

LENAERTS, KOENRAAD, PIET VAN NUFFEL, and ROBERT BRAY (editor) (2005) *Constitutional Law of the European Union* (London: Sweet & Maxwell).

LENEL, HANS OTTO (1989) 'Evolution of the Social Market Economy' in Peacock, Alan and Hans Willgerodt (eds) *German Neo-Liberals and the Social Market Economy* (London: Macmillan).

LERNER, A P (1934) 'The Concept of Monopoly and the Measurement of Monopoly Power' 1 *Rev Econ Stud* 157–175.

LETWIN, WILLIAM (1981) *Law and Economic Policy in America: The Evolution of the Sherman Antitrust Act* (Phoenix Edition Chicago: University of Chicago Press).

LEVER, JEREMY, and SILKE NEUBAUER (2000) 'Vertical Restraints, Their Motivation and Justification' 21 *ECLR* 7–23.

——(2002) 'The German Monopolies Commission's Report on the Problems Consequent Upon the Reform of the European Cartel Procedures' 23 *ECLR* 321–325.

LEVER QC, JEREMY, and GEORGE PERETZ (1999) 'Comments on the EC Commission's White Paper on Modernisation of the Rules Implementing Articles 81(ex 85) and 82 (ex 86) of the EC Treaty' *Written Evidence Submitted to The Select Committee appointed to consider European Union documents and other matters relating to the European Union Sub-Committee E (Law and Institutions) Fourth Report, 15 February 2000.*

LEVY, NICHOLAS (2003) *European Merger Control Law: A Guide to the Merger Regulation* (Release 1 2004 Newark, NJ: LexisNexis).

LEWIS, DAVID K (1969) *Convention: A Philosophical Study* (Cambridge, Mass.: Harvard University Press).

LIDGARD, HANS HENRIK (1997) 'Unilateral Refusal to Supply: An Agreement in Disguise?' 18 *ECLR* 352–360.

LIEBELER, WESLEY J (1982) 'Intrabrand 'Cartels' under GTE Sylvania' 30 *UCLA L Rev* 1–51.

——(1983) '1983 Economic Review of Antitrust Developments: The Distinction between Price and Nonprice Distribution Restrictions' 31 *UCLA L Rev* 384–422.

LIEBERMAN, M B (1987) 'Excess Capacity as a Barrier to Entry: An Empirical Appraisal' 35 *J Ind Econ* 607–627.

LIPSEY, RICHARD G, PAUL N COURANT, DOUGLAS D PURVIS, and PETER O STEINER (1993) *Economics* (Tenth Edition New York: HarperCollins College Publishers).

LONBAY, JULIAN (1996) 'Case C-55/94 Reinhard Gebhard v. Consiglio Dell'ordine Degli Avvocati e Procuratori di Milano' 33 *CML Rev* 1073–1087.

London Economics (March 1994) Barriers to Entry and Exit in UK Competition Policy *Office Of Fair Trading* Report 2.

LOUGHLIN, MARTIN (2003) *The Idea of Public Law* (Oxford: Oxford University Press).

LOURI, VICTORIA (2002) 'Undertaking' as a Jurisdictional Element for the Application of EC Competition Rules' 29 *LIEI* 143–176.

LUTZ, FREIEDRICH A (1989) 'Observations on the Problem of Monopolies' in Peacock, Alan and Hans Willgerodt (eds) *Germany's Social Market Economy: Origins and Evolution* (Basingstoke: Macmillan).

MADURO, MIGUEL POIARES (1997) 'Reforming the Market or the State? Article 30 and the European Constitution: Economic Freedom and Political Rights' 3 *ELJ* 55–82.

MAHER, IMELDA (2000) 'Re-Imagining the Story of European Competition Law' 20 *OJLS* 155–166.

MANN, F A (1973) 'The Dyestuffs Case in the Court of Justice of the European Communities' 22 *Int'l & Comp LQ* 35–50.

MANZINI, PIETRO (2002) 'The European Rule of Reason—Crossing the Sea of Doubt' 23 *ECLR* 392–399.

MARENCO, GIULIANO (1987) 'Competition between National Economies and Competition between Businesses—A Response to Judge Pescatore' 10 *Fordham Int'l LJ* 420–443.

MARKOVITS, RICHARD S (1975a) 'Some Preliminary Notes on the American Antitrust Laws' Economic Tests of Legality' 27 *Stan L Rev* 841–858.

——(1975b) 'A Basic Structure from Microeconomic Policy Analysis in Our Worse-Than-Second-Best World: A Proposal and Related Critique of the Chicago Approach to the Study of Law and Economics' *Wis L Rev* 950–1080.

MARSDEN, PHILIP B (1997) 'Inducing Member State Enforcement of European Competition Law: A Competition Policy Approach to Antitrust Federalism' 18 *ECLR* 234–241.

MARTIN, STEPHEN (2001) *Industrial Organization: A European Perspective* (Oxford: Oxford University Press).

MASON, EDWARD S (1937) 'Monopoly in Law and Economics' 34 *Yale LJ* 34–49.

MASSEY, PATRICK (1998) 'The Treatment of Vertical Restraints under Competition Law' *Irish Competition Authority Discussion Paper No 4* available at http://www.tca.ie/discpap.html.

MASTROMANOLIS, EMMANUEL P (1995) 'Insights from US Antitrust Law on Exclusive and Restricted Territorial Distribution: The Creation of A New Legal Standard for European Union Competition Law' 15 *University of Pennsylvania Journal of International Business Law* 559–647.

MATHEWSON, G FRANK, and RALPH A WINTER (1987) 'The Competitive Effects of Vertical Agreements: Comment' 77 *Amer Econ Rev* 1057–1062.

MATTE QC, FRANCINE (1998) 'Competition Policy Objectives' in Ehlermann, Claus Dieter and Laraine L Laudati (eds) *European Competition Law Annual 1997: The Objectives of Competition Policy* (Oxford: Hart) 85–109.

MAURICE, S CHARLES, and OWEN R PHILLIPS (1986) *Economic Analysis: Theory and Application* (Fifth Edition Homewood, Illinois: Irwin).

MAVROIDIS, PETROS C, and DAMIEN J NEVEN (2001) 'From the White Paper to the Proposal for a Council Regulation: How to Treat the New Kids on the Block?' 28 *LIEI* 151–171.

MCGEE, JOHN S (1971) *In Defense of Industrial Concentration* (New York: Praeger).

MCLEOD, A N, and F H HAHN (1949) 'Proportionality, Divisibility, and Economies of Scale: Two Comments' 63 *Quart J Econ* 128–137.

McNutt, Patrick (2000) 'The Appraisal of Vertical Agreements: Competitiveness, Efficiency, Competitive Harm and Dynamic Conduct' in Rivas, José and Margot Horspool (eds) *Modernisation and Decentralisation of EC Competition Law* (The Hague: Kluwer Law International) 49–68.

Meese, Alan J (2000) 'Farewell to the Quick Look: Redefining the Scope and Content of the Rule of Reason' 68 *Antitrust LJ* 461–498.

Meinhardt, Marcel, and Astrid Waser (2005) 'The Revised Swiss Competition Law: Direct Fines, Leniency and Notification' 26 *ECLR* 349–354.

Menendez, Augustin Jose (2003) 'The Sinews of Peace: Rights to Solidarity in the Charter of Fundamental Rights of the European Union' 16 *Ratio Juris* 374–398.

Milgrom, P, and J Roberts (1982) 'Predation, Reputation, and Entry Deterrence' 27 *J Econ Theory* 280–312.

Milner-Moore, G R 'The Accountability of Private Parties under the Free Movement of Goods Principle' *Harvard Jean Monnet Working Paper No 9/95* http://www.jeanmonnetprogram.org/papers/95/9509ind.html.

Mitchell, Gregory (2002) 'Taking Behavioralism Too Seriously? The Unwarranted Pessimism of the New Behavioral Analysis of Law' 43 *William and Mary Law Review* 1907–2021.

Molle, Willem (2001) *The Economics of European Integration: Theory, Practice, Policy* (Fourth Edition Aldershot: Ashgate).

Monbiot, George (2000) *Captive State: The Corporate Takeover of Britain* (London: Pan).

Monopolkommission (2000) *Cartel Policy Change in the European Union?: On the European Commission's White Paper of 28th April 1999; Special Report by the German Monopolies Commission Pursuant to Sec. 44, Para. 1 of the Act against Restraints of Competition (Gwb)* (Baden-Baden: Nomos).

Montag, Frank (1996) 'The Case for a Radical Reform of the Infringement Procedure under Regulation 17' 17 *ECLR* 428–437.

Montana, Laura, and Jane Jellis (2003) 'The Concept of Undertaking in EC Competition Law and Its Application to Public Bodies: Can You Buy Your Way into Article 82?' 2 *Competition Law Journal* 110–118.

Monti, Giorgio (2002) 'Article 81 EC and Public Policy' 39 *CML Rev* 1057–1099.

Monti, Mario (2000) 'The Application of Community Competition Law by the National Courts' SPEECH/00/466.

——(2002) 'Competition and the Consumer: What Are the Aims of European Competition Policy' SPEECH/02/79.

Morris, Derek (2002) 'The Enterprise Act: Aspects of the New Regime' 22 *Economic Affairs* 15–24.

Mortelmans, Kamiel (2001) 'Towards Convergence in the Application of the Rules on Free Movement and on Competition?' 38 *CML Rev* 613–649.

Möschel, Wernard (1989) 'Competition Policy from an Ordo Point of View' in Peacock, Alan and Hans Willgerodt (eds) *German Neo-Liberals and the Social Market Economy* (London: Macmillan).

MOTTA, MASSIMO (2004) *Competition Policy: Theory and Practice* (Cambridge: Cambridge University Press).

MÜLLER-ARMACK, ALFRED (1989) 'The Meaning of the Social Market Economy' in Peacock, Alan and Hans Willgerodt (eds) *Germany's Social Market Economy: Origins and Evolution* (Basingstoke: Macmillan).

MURIS, TIMOTHY J (1997) 'Economics and Antitrust' 5 *Geo Mason L Rev* 303–312.

——(1998) 'The Federal Trade Commission and the Rule of Reason: In Defense of Massachusetts Board' 66 *Antitrust LJ* 773–804.

National Economic Research Associates (1992) *Market Definition in UK Competition Policy* Office Of Fair Trading Research Paper 1.

NAZERALI, JULIE, and DAVID COWAN (2000) 'Unlocking EU Distribution Rules—Has the European Commission Found the Right Keys?' 21 *ECLR* 50–56.

NEALIS, PETER (2000) 'Per Se Legality: A New Standard in Antitrust Adjudication under the Rule of Reason' 61 *Ohio St LJ* 347–398.

NELSON, RICHARD R, and SIDNEY G WINTER (1982) 'The Schumpeterian Tradeoff Revisited' 72 *Amer Econ Rev* 114–132.

NEVEN, DAMIEN (1998) 'Competition Policy Objectives' in Ehlermann, Claus Dieter and Laraine L Laudati (eds) *European Competition Law Annual 1997: The Objectives of Competition Policy* (Oxford: Hart) 111–118.

——, PENELOPE PAPANDROPOULOS, and PAUL SEABRIGHT (1998) *Trawling for Minnows: European Competition Policy and Agreements between Firms* (London: CEPR).

NICHOLLS, A J (1994) *Freedom with Responsibility: The Social Market Economy in Germany, 1918–1963* (Oxford: Clarendon Press).

NICOLAIDES, PHEDON (2000) 'An Essay on Economics and the Competition Law of the European Community' 27 LIEI 7–24.

NIHOUL, P L G (2004) 'Authorities, Competition and Electronic Communication: Towards Institutional Competition in the Information Society' in Graham, Cosmo and Fiona Smith (eds) *Competition, Regulation and the New Economy* (Oxford: Hart) 91–124.

NIHOUL, PAUL (2002) 'Do Workers Constitute Undertakings for the Purpose of the Competition Rules?' 25 *ELRev* 408–414.

NORRIE, ALAN W (1989) 'Oblique Intention and Legal Politics' *CrimLR* 793–807.

Note (1993) 'IAZ International Belgium v Commission' 4 *ECLR* 248–251.

——(2003) 'Economic Activity (Sickness Funds) the AOK Case' 26 *CompLaw EC* 136–138.

——(2004) 'The EC Judgment in AOK: Can a Major Public Sector Purchaser Control the Prices It Pays or Is It Subject to the Competition Act? Cases C-264/01, C-306/01, C-354/01, and C-355/01: AOK Bundesverband v Ichthyol, ECJ March 16, 2004' *Public Procurement Law Review* NA95-NA97.

NOURRY, ALEX, and TRUDI LODGE (1997) 'European Commission Green Paper on Vertical Restraints in EC Competition Policy' 8 *ICCLR* 191–195.

O'LOUGHLIN, ROSEMARY (2003) 'EC Competition Rules and Free Movement Rules: An Examination of the Parallels and Their Furtherance by the ECJ Wouters Decision' 24 *ECLR* 62–69.

Oberdorfer, Conrad W, and Alfred Gleiss (1963) *Common Market Cartel Law* (Third Edition New York: Commerce Clearing House).

Odudu, Okeoghene (2001a) 'Interpreting Article 81(1): Object as Subjective Intention' 26 *ELRev* 60–75.

——(2001b) 'Interpreting Article 81(1): The Object Requirement Revisited' 26 *ELRev* 379–390.

——(2002a) 'A New Economic Approach to Article 81(1)' 27 *ELRev* 100–105.

——(2002b) 'The Role of Specific Intent in Section 1 of the Sherman Act: A Market Power Test?' 25 *W Comp* 463–491.

Office of Fair Trading (2004a) The Competition Act 1998 and Public Bodies OFT 443.

——(2004b) Assessment of Market Power OFT 415.

——(2004c) OFT's Guidance as to the Appropriate Amount of a Penalty: Understanding Competition Law OFT 423.

——(2004d) Market Definition OFT 403.

——(2005) Leniency in Cartel Cases: A Guide to the Leniency Programme for Cartels OFT436.

Oliver, Dawn (1999) *Common Values and the Public-Private Divide* (London: Butterworths).

——(2000) 'The Frontiers of the State: Public Authorities and Public Functions under the Human Rights Act' *Public Law* 476–493.

Oliver Jr, Henry M (1960) 'German Neoliberalism' 74 *Quart J Econ* 117–149.

Oliver, Peter (1999) 'Some Further Reflections on the Scope of Articles 28–30 (ex 30–36) EC' 36 *CML Rev* 783–806.

—— and Malcolm A Jarvis (2003) Free Movement of Goods in the European Community: Under Articles 28 to 30 of the EC Treaty (Fourth Edition London: Sweet & Maxwell).

Organisation for Economic Co-operation and Development (1997) 'Judicial Enforcement of Competition Law' OCDE/GD(97)200.

——(1999) 'Complementarities between Trade and Competition Policies' COM/TD/DAFFE/CLP(98)98/Final.

Ortiz Blanco, Luis (1996) *European Community Competition Procedure* (Oxford: Clarendon Press).

Oswalt-Eucken, Irene (1994) 'Freedom and Economic Power: Neglected Aspects of Walter Eucken's Work' 21 *J Econ Stud* 38–45.

Panel Discussion (1998) 'Future Competition Law' in Ehlermann, Claus Dieter and Laraine L Laudati (eds) *European Competition Law Annual 1997: The Objectives of Competition Policy* (Oxford: Hart Publishing) 459–492.

Panzar, John C, and Robert D Willig (1981) 'Economies of Scope' 71 *Amer Econ Rev* 268–272.

Papandreou, Andreas G (1949) 'Market Structure and Monopoly Power' 39 *Amer Econ Rev* 883–897.

Patterson, Mark R (1997) 'Coercion, Deception, and Other Demand-Increasing Practices in Antitrust Law' 66 *Antitrust LJ* 1–90.

——(2000) 'The Role of Market Power in the Rule of Reason' 68 *Antitrust LJ* 429–460.

PAULWEBER, MICHAEL (2000) 'The End of a Success Story? The European Commission's White Paper on the Modernisation of the European Competition Law: A Comparative Study About the Role of the Notification of Restrictive Practices within the European Competition and the American Antitrust Law' *23 W Comp* 3–48.

PAWSON, RAY, and NICHOLAS TILLEY (1997) *Realistic Evaluation* (London: Sage).

PEACOCK, ALAN, and HANS WILLGERODT (1989a) 'German Liberalism and Economic Revival' in Peacock, Alan and Hans Willgerodt (eds) *Germany's Social Market Economy: Origins and Evolution* (Basingstoke: Macmillan).

——(1989b) 'Overall View of the German Liberal Movement' in Peacock, Alan and Hans Willgerodt (eds) *German Neo-Liberals and the Social Market Economy* (London: Macmillan).

PEEPERKORN, LUC (1998) 'The Economics of Verticals' June *EC Competition Policy Newsletter* 10–17.

——(2002) 'EC Vertical Restraints Guidelines: Effects-Based or Per Se Policy?—A Reply' 23 *ECLR* 38.

PEETERS, JAN (1989) 'The Rule of Reason Revisited: Prohibition on Restraints of Competition in the Sherman Act and the EEC Treaty' 37 *Am J Comp* L 521–570.

PERITZ, RUDOLPH J (1989) 'The "Rule of Reason" in Antitrust Law: Property Logic in Restraint of Competition' 40 *Hastings LJ* 285–342.

PERSKY, JOSEPH (1995) 'Retrospectives: The Ethology of Homo Economicus' 9 *J Econ Perspect* 221–231.

PESCATORE, PIERRE (1983) 'The Doctrine of "Direct Effect": An Infant Disease of Community Law' 8 *ELRev* 155–177.

——(1987) 'Public and Private Aspects of European Community Competition Law' 10 *Fordham Int'l LJ* 373–419.

PETERSMANN, ERNST-ULRICH (1999) 'Legal, Economic and Political Objectives of National and International Competition Policies: Constitutional Functions of WTO 'Linking Principles' for Trade and Competition' 34 *New Eng L Rev* 145–162.

PHEASANT, JOHN, and DANIEL WESTON (1997) 'Vertical Restraints, Foreclosure and Article 85: Developing an Analytical Framework' 18 *ECLR* 323–328.

PHLIPS, LOUIS (1995) *Competition Policy: A Game-Theoretic Perspective* (Cambridge: Cambridge University Press).

PIRAINO JR, THOMAS A (1991) 'Reconciling the Per Se and the Rule of Reason Approaches to Antitrust Analysis' 64 *S Cal L Rev* 685–739.

——(1994) 'Making Sense of the Rule of Reason: A New Standard for Section 1 of the Sherman Act' 47 *Vand L Rev* 1753–1805.

PITOFSKY, ROBERT (1979) 'The Political Content of Antitrust' 127 *U Pa L Rev* 1051.

——(1995) 'Antitrust Modified: Education, Defense, and Other Worthy Enterprises' 9 *Antitrust* 23–25.

——(1989) *An Introduction to Law and Economics* (Second Edition Boston: Little Brown).

POLINSKY, A MITCHELL (1989) *An Introduction to Law and Economics* (Second Edition Boston: Little Brown)
POPPER, KARL (2002) *The Logic of Scientific Discovery* (London: Routledge Classics).
PORTER, MICHAEL E (2001) 'Competition and Antitrust: Toward a Productivity-Based Approach to Evaluating Mergers and Joint Ventures' 46 *Antitrust Bull* 919–957.
POSNER, RICHARD A (1969) 'Oligopoly and the Antitrust Laws: A Suggested Approach' 21 *Stan L Rev* 1562–1606.
——(1970) 'Antitrust Policy and the Consumer Movement' 15 *Antitrust Bull* 361–366.
——(1975) 'The Social Costs of Monopoly and Regulation' 83 *J Polit Economy* 807–828.
——(1976) *Antitrust Law: An Economic Perspective* (Chicago: University of Chicago Press).
——(1977) 'The Rule of Reason and the Economic Approach: Reflections on the Sylvania Decision' 45 *U Chi L Rev* 1–20.
——(1979a) 'The Chicago School of Antitrust Analysis' 127 *U Pa L Rev* 925–948.
——(1979b) 'Information and Antitrust: Reflections on the Gypsum and Engineers Decisions' 67 *Geo LJ* 1187–1203.
——(1981a) *The Economics of Justice* (Cambridge, Massachusetts: Harvard University Press).
——(1981b) 'The Next Step in the Antitrust Treatment of Restricted Distribution: Per Se Legality' 48 *U Chi L Rev* 6–26.
——(1990) *The Problems of Jurisprudence* (Cambridge, Mass: Harvard University Press).
——(2001) *Antitrust Law* (Second Edition Chicago: University of Chicago Press).
——(2003) *Economic Analysis of Law* (Sixth Edition New York: Aspen Publishers).
PRECHAL, SACHA (2000) 'Does Direct Effect Still Matter?' 37 *CML Rev* 1047–1069.
PRICE, RICHARD G (1989) 'Market Power and Monopoly Power in Antitrust Analysis' 75 *Cornell L Rev* 190–217.
PRINSSEN, JOLANDE M, and ANNETTE SCHRAUWEN, eds. (2002) *Direct Effect: Rethinking a Classic of EC Legal Doctrine* (Groningen: Europa Law Publishing).
PROSSER, TONY (2005) *The Limits of Competition Law: Markets and Public Services* (Oxford: Oxford University Press).
QUAID, JENNIFER A (1998) 'The Assessment of Corporate Criminal Liability on the Basis of Corporate Identity: An Analysis' 43 *McGill LJ* 67–114.
QUINN, MARY, and NICHOLAS MACGOWAN (1987) 'Could Article 30 Impose Obligations on Individuals' 12 *ELRev* 163–178.
RAPP, RICHARD T (1995) 'The Misapplication of the Innovation Market Approach to Merger Analysis' 64 *Antitrust LJ* 19–47.
RASMUSSEN, HJALTE (1986) *On Law and Policy in the European Court of Justice: A Comparative Study in Judicial Policymaking* (Dordrecht: Nijhoff).
RAZ, JOSEPH (1979) *The Authority of Law: Essays on Law and Morality* (Oxford: Clarendon).

Reeves, Tony, and Philip Brentford (2000) 'A New Competition Policy for a New Millennium?' 11 *ICCLR* 75–81.

Reich, Norbert (2005) 'The "Courage" Doctrine Encouraging or Discouraging Compensation for Antitrust Injuries?' 42 *CML Rev* 35–66.

Rennie, Jane (2005) 'The "ACCC Leniency Policy for Cartel Conduct"—Australia's Role in the War on Collusion' 20 *Journal of International Banking Law and Regulation* 165–182.

Riley, Alan J (1998) 'Vertical Restraints: A Revolution?' 19 *ECLR* 483–492.

Ritter, Lennart, W David Braun, and Francis Rawlinson (2000) *EC Competition Law: A Practitioner's Guide* (Second Edition The Hague: Kluwer Law International).

Rivas, José, and Jonathan Branton (2003) 'Developments in EC Competition Law in 2002: An Overview' 40 *CML Rev* 1187–1240.

Robertson, Aidan (2002) 'Professional Rules under the Competition Act 1998' 1 *Competition Law Journal* 93–100.

Rodger, Barry J, and Stuart R Wylie (1997) 'Taking the Community Interest Line: Decentralisation and Subsidiarity in Competition Law Enforcement' 18 *ECLR* 485–491.

—— and Angus MacCulloch (1999) *Competition Law and Policy in the EC and UK* (London: Cavendish).

Romer, Paul M (1986) 'Increasing Returns and Long Run-Growth' 94 *J Polit Economy* 1002–1037.

——(1996) 'Why, Indeed, in America? Theory, History, and the Origins of Modern Economic Growth' 86 *Amer Econ Rev* 202–206.

Röpke, Wilhelm (1960) *A Humane Economy: The Social Framework of the Free Market* (South Bend, Indiana: Gateway Editions).

Rose, Vivien, ed. (1993) *Bellamy & Child: Common Market Law of Competition* (Fourth Edition London: Sweet & Maxwell).

Rose-Ackerman, Susan (1992) *Rethinking the Progressive Agenda: The Reform of the American Regulatory State* (New York: The Free Press).

Ross, Stephen F (1993) *Principles of Antitrust Law* (Westbury, NY: Foundation Press).

Ross, Thomas W (2004) 'Viewpoint: Canadian Competition Policy: Progress and Prospects' 37 *Canadian Journal of Economics* 243–268.

Rossi, Peter H, and James D Wright (1984) 'Evaluation Research: An Assesment' 10 *Annual Review of Sociology* 331–352.

Rostow, Eugene V (1947) 'The New Sherman Act: A Positive Instrument of Progress' 14 *U Chi L Rev* 567–600.

Roth QC, ed. (2001) *Bellamy & Child: European Community Law of Competition* (Fifth Edition London: Sweet & Maxwell).

Russell, Bertrand (1980) *The Problems of Philosophy* (Ninth Impression with Appendix Oxford: Oxford University Press).

Sainsbury, Mark (1995) 'Philosophical Logic' in Grayling, A C (ed) *Philosophy: A Guide through the Subject* (Oxford: Oxford University Press).

Salop, Steven C (1979) 'Strategic Entry Deterrence' 69 *Amer Econ Rev* 335–338.

SALOP, STEVEN C and DAVID T SCHEFFMAN (1983) 'Raising Rivals' Costs' 73 *Amer Econ Rev* 267–271.

——(2000) 'The First Principles Approach to Antitrust, Kodak, and Antitrust at the Millennium' 68 *Antitrust LJ* 187–202.

—— ed. (1981) *Strategy, Predation, and Antitrust Analysis* (Washington, DC: Federal Trade Commission Bureau of Economics Bureau of Competition).

SAVIN, JAMES ROB (1997) 'Tunney Act' 96: Two Decades of Judicial Misapplication' 46 *Emory LJ* 363–387.

SCHAUB, ALEXANDER (1998) 'Competition Policy Objectives' in Ehlermann, Claus Dieter and Laraine L Laudati (eds) *European Competition Law Annual 1997: The Objectives of Competition Policy* (Oxford: Hart) 119–128.

——(1999) 'EC Competition System: Proposals for Reform' in Hawk, Barry E (ed) *Annual Proceedings of the Fordham Corporate Law Institute: International Antitrust Law & Policy 1998* (New York, NY: Juris Publishing) 129–156.

SCHEPEL, HARM (2002) 'Delegation of Regulatory Powers to Private Parties under EC Competition Law: Towards a Procedural Public Interest Test' 39 *CML Rev* 31–51.

SCHERER, F M (1987) 'Antitrust, Efficiency, and Progress' 62 *NYU L Rev* 998–1019.

—— and DAVID ROSS (1990) *Industrial Market Structure and Economic Performance* (Third Edition Boston: Houghton Mifflin).

——(1992) 'Schumpeter and Plausible Capitalism' 30 *J Econ Lit* 1416–1433.

——(2001) 'Some Principles for Post-Chicago Antitrust Analysis' 52 *Case W Res L Rev* 5–23.

SCHMALENSEE, RICHARD (1973) 'A Note on the Theory of Vertical Integration' 81 *J Polit Economy* 442–449.

——(1982a) 'Another Look at Market Power' 95 *Harv L Rev* 1789–1816.

——(1982b) 'Product Differentiation Advantages of Pioneering Brands' 72 *Amer Econ Rev* 349–365.

——(1982c) 'Antitrust and the New Industrial Economics' 72 *Amer Econ Rev* 24–28.

——(1983) 'Product Differentiation Advantages of Pioneering Brands: Errata' 73 *Amer Econ Rev* 250.

SCHMID, CHRISTOPH U (2000) 'Diagonal Competence Conflicts between European Competition Law and National Regulation—A Conflict of Laws Reconstruction of the Dispute on Book Price Fixing' 8 *ERPL* 155–172.

SCHUMPETER, JOSEPH A (1975) *Capitalism, Socialism and Democracy* (First Harper Colophon Edition; with a new introduction by Tom Bottomore New York: Harper & Row).

SCHWARTZ, ALAN, and ROBERT E SCOTT (2003) 'Contract Theory and the Limits of Contract Law' 113 *Yale LJ* 541–620.

SCHWARTZ, MARIUS (1987) 'The Competitive Effects of Vertical Agreements: Comment' 77 *Amer Econ Rev* 1063–1068.

SCITOVSKY, TIBOR (1956) 'Economies of Scale, Competition, and European Integration' 46 *Amer Econ Rev* 71–91.

SEN, AMARTYA (1993) 'Markets and Freedoms: Achievements and Limitations of the Market Mechanism in Promoting Individual Freedoms' 45 *Oxford Econ Pap* 519–541.

SHAVELL, STEVEN (2004) *Foundations of Economic Analysis of Law* (Cambridge, Mass.: Belknap).

SHAW, JOSEPHINE (1988) 'A Healthy Monopoly for a Dying Trade?' 13 *ELRev* 422–426.

SILBERSTON, AUBREY (1972) 'Economies of Scale in Theory and Practice' 82 *Econ J* 369–391.

SINGER, EUGENE M (1968) *Antitrust Economics: Selected Legal Cases and Economic Models* (Englewood Cliffs, New Jersey: Prentice-Hall).

SIRAGUSA, MARIO (1998) 'The Millennium Approaches: Rethinking Article 85 and the Problems and Challenges in the Design and Enforcement of the EC Competition Rules' 21 *Fordham Int'l LJ* 650–678.

——(2000) 'A Critical Review of the White Paper on the Reform of the EC Competition Law Enforcement Rules' 23 *Fordham Int'l LJ* 1089–1127.

SKILBECK, JENNIFER (2002) 'Bettercare: The Conflict between Social Policy and Economic Efficiency' 1 *Competition Law Journal* 260–269.

SLAPPER, GARY, and STEVE TOMBS (1999) *Corporate Crime* (Harlow: Longman).

SLOT, PIET JAN (1987) 'The Application of Articles 3(F), 5 and 85 to 94 EEC' 12 *ELRev* 179–189.

——(1993) 'Introduction' in Slot, Piet Jan and Alison McDonnell (eds) *Procedure and Enforcement in EC and US Competition Law* (London: Sweet & Maxwell) ix-xvii.

SMITH, MARTIN (2001) *Competition Law: Enforcement and Procedure* (London: Butterworths).

SMITH, STEPHEN A (2004) *Contract Theory* (Oxford: Oxford University Press).

SNELL, JUKKA (2002) 'Private Parties and the Free Movement of Goods and Services' in Andenas, Mads and Wulf-Henning Roth (eds) *Services and Free Movement in EU Law* (Oxford: Oxford University Press) 211–243.

——(2003) 'Who's Got the Power? Free Movement and Allocation of Competences in EC Law' 22 *YEL* 323–351.

SNYDER, FRANCIS (1990) *New Directions in European Community Law* (London: Weidenfeld and Nicolson).

SOAMES, TREVOR (1996) 'An Analysis of the Principles of Concerted Practice and Collective Dominance: A Distinction without a Difference' 17 *ECLR* 24–39.

SPAVENTA, ELEANOR (2004a) 'Public Services and European Law: Looking for Boundaries' in Bell, John and Alan Dashwood (eds) *Cambridge Yearbook of European Legal Studies 2003* (Oxford: Hart) 271–291.

——(2004b) 'From Gebhard to Carpenter: Towards a (Non-)Economic European Constitution' 41 *CML Rev* 743–773.

SPENCE, A M (1977) 'Entry, Capacity, Investment and Oligopolistic Pricing' 8 *Bell J Econ* 534–544.

SQUIRE, LYN (1980) 'On the Use of Distributional Weights in Social Cost-Benefit Analysis' 88 *J Polit Economy* 1048–1049.

STEINDORFF, ERNST (1972) 'Annotation on the Decision of the European Court in the Dyestuff Cases of July 14, 1972' 9 *CML Rev* 502–510.

STELZER, IRWIN M (2002) 'Innovation, Price Competition and the "Tilt" of Competition Policy' 22 *Economic Affairs* 25–31.

STEVENS, ROBERT BOCKING, and BASIL S YAMEY (1965) *The Restrictive Practices Court: A Study of the Judicial Process and Economic Policy* (London: Weidenfeld & Nicolson).

STIGLER, GEORGE JOSEPH (1954) 'The Economist Plays with Blocs' 44 *Amer Econ Rev* 7–14.

——(1957) 'Perfect Competition, Historically Contemplated' 65 *J Polit Economy* 1–17.

——(1976) 'The Xistence of X-Efficiency' 66 *Amer Econ Rev* 213–216.

——(1982) 'The Economists and the Problem of Monopoly' 72 *Amer Econ Rev* 1–11.

——(1983a) 'The Economic Effects of the Antitrust Laws' in Stigler, George J (ed) *The Organization of Industry* (Chicago: The University of Chicago Press) 259–295.

——(1983b) 'A Theory of Oligopoly' in Stigler, George J (ed) *The Organization of Industry* (Chicago: The University of Chicago Press) 39–63.

——(1983c) 'Barriers to Entry, Economies of Scale, and Firm Size' in Stigler, George Joseph (ed) *The Organization of Industry* (Chicago: University of Chicago Press).

STIGLITZ, JOSEPH E (2000) *Economics of the Public Sector* (Third Edition London: W W Norton).

—— and JOHN DRIFFILL (2000) *Economics* (London: W W Norton).

STOCKING, GEORGE W, and MYRON WATKINS (1946) *Cartels in Action* (New York: The Twentieth Century Fund).

STURGEON, SCOTT (1995) 'Knowledge' in Grayling, A C (ed) *Philosophy: A Guide through the Subject* (Oxford: Oxford University Press).

SUBIOTTO, ROMANO, and FILIPPO AMATO (2000) 'Preliminary Analysis of the Commission's Reform Concerning Vertical Restraints' 23 *W Comp* 5–26.

——(2001) 'The Reform of the European Competition Policy Concerning Vertical Restraints' 69 *Antitrust LJ* 147–194.

SULLIVAN, LAWRENCE A, and WARREN S GRIMES (2000) *The Law of Antitrust: An Integrated Handbook* (St. Paul, MN: West Group).

SULLIVAN, LAWRENCE ANTHONY (1977a) 'Economics and More Humanistic Disciplines: What Are the Sources of Wisdom for Antitrust' 125 *U Pa L Rev* 1214–1243.

——(1977b) *Handbook of the Law of Antitrust* (St. Paul: West Publishing Company).

SWANN, DENNIS (2000) *The Economics of Europe: From Common Market to European Union* (Ninth Edition London: Penguin).

SZYSZCZAK, ERIKA (2004) 'Public Services in the New Economy' in Graham, Cosmo and Fiona Smith (eds) *Competition, Regulation and the New Economy* (Oxford: Hart) 185–206.

TELSER, LESTER G (1960) 'Why Should Manufacturers Want Fair Trade?' 3 *J Law Econ* 86–105.

TEMPLE LANG, JOHN (1981) 'Community Antitrust Law—Compliance and Enforcement' 18 *CML Rev* 335–362.

TEMPLE LANG, JOHN (2004) 'National Measures Restricting Competition and National Authorities under Article 10 EC' 29 *ELRev* 397–406.

TERHORST, GEORG (2000) 'The Reformation of the EC Competition Policy on Vertical Restraints' 21 *Nw J Int'l L & Bus* 343–378.

TIROLE, JEAN (1988) *The Theory of Industrial Organization* (Twelfth Printing 2001 Cambridge, Massachusetts: MIT Press).

TOEPKE, UTZ PETER (1982) *EEC Competition Law: Business Issues and Legal Principles in Common Market Antitrust Cases* (New York: Wiley).

TORRE, ANDREW (2005) 'The Use of Own Price Elasticity of Demand in Competition Law' 26 *ECLR* 468–473.

TREITEL, G H (2004) *An Outline of the Law of Contract* (Oxford: Oxford University Press).

TULLOCK, GORDON (1971) *The Logic of the Law* (New York: Basic Books).

TURNER, DONALD F (1962) 'The Definition of Agreement under the Sherman Act: Conscious Parallelism and Refusals to Deal' 75 *Harv L Rev* 655–706.

——(1969) 'The Scope of Antitrust and Other Economic Regulatory Policies' 82 *Harv L Rev* 1207–1244.

VAN DEN BERGH, ROGER (1996) 'Modern Industrial Organisation Versus Old-Fashioned European Competition Law' 17 *ECLR* 75–87.

—— and PETER D CAMESASCA (2000) 'Irreconcilable Principles? The Court of Justice Exempts Collective Labour Agreements from the Wrath of Antitrust' 25 *ELRev* 492–508.

————(2001) *European Competition Law and Economics: A Comparative Perspective* (Antwerpen: Intersentia).

VAN DEN BOGAERT, STEFAAN (2000) 'The Court of Justice on the Tatami: Ippon, Waza-Ari or Koka?' 25 *ELRev* 554–563.

——(2002) 'Horizontality: The Court Attacks?' in Barnard, Catherine and Joanne Scott (eds) *The Law of the Single European Market* (Oxford: Hart) 123–152.

VAN GERVEN, GERWIN, and EDURNE NAVARRO VARONA (1994) 'The Wood Pulp Case and the Future of Concerted Practices' 31 *CML Rev* 575–608.

VAN MIERT, KARL (1998) 'The Future of European Competition Policy' Presentation of the Ludwig Erhard Prize http://europa.eu.int/comm/competition/speeches/text/sp1998_042_en.html.

VARONA, EDURNE NAVARRO, ANDRES FOTNT GALARZA, JAIME FOLGUERA CRESPO, and JUAN BRIONES ALONSO (2002) *Merger Control in the European Union: Law, Economics and Practice* (Oxford: Oxford University Press).

VELJANOVSKI, CENTO G (1982) *The New Law and Economics: A Research Review* (Oxford: Centre for Socio-Legal Studies Wolfson College).

VENIT, JAMES S (2003) 'Brave New World: The Modernization and Decentralization of Enforcement under Articles 81 and 82 of the EC Treaty' 40 *CML Rev* 545–580.

VERNON, JOHN M, and DANIEL A GRAHAM (1971) 'Profitability of Monopolization by Vertical Integration' 79 *J Polit Economy* 924–925.

VEROUDEN, VINCENT (2003) 'Vertical Agreements and Article 81 (1) EC: The Evolving Role of Economic Analysis' 71 *Antitrust LJ* 525–576.

VICKERS, JOHN (1995) 'Concepts of Competition' 47 *Oxford Econ Pap* 1–23.

VISCUSI, W KIP, JOHN M VERNON, and JOSEPH E HARRINGTON Jr (2000) *Economics of Regulation and Antitrust* (Third Edition Cambridge, Massachusetts: MIT Press).

VOSSESTEIN, ADRIAN J (2002) 'Case C-35/99, Arduino, Judgment of 19 February 2000, Full Court; Case C-309/99, Wouters et al v Algemene Raad Van de Nederlandse Orde Van Advocaten, Judgment of 19 February 2002, Full Court; Not yet Reported' 39 *CML Rev* 841–863.

WAELBROECK, MICHEL, and ALDO FRIGNANI (1999) *European Competition Law* (Ardsley, NY: Transnational Publishers).

WALLER, SPENCER WEBER (2001) 'The Language of Law and the Language of Business' 52 *Case W Res L Rev* 283–338.

WARD, ANGELA (2003) 'Locus Standi under Article 230(4) of the EC Treaty: Crafting a Coherent Test for A "Wobbly Polity"' 22 *YEL* 45–77.

WARREN-BOULTON, FREDERICK R (1974) 'Vertical Control with Variable Proportions' 82 *J Polit Economy* 783–802.

WEATHERILL, STEPHEN (1989) 'Discrimination on Grounds of Nationality in Sport' 9 *YEL* 55–92.

——(1996) 'Case C-415/93, Union Royale Belge des Societes de Football Association Asbl v. Jean-Marc Bosman; Royal Club Liegois SA v. Jean-Marc Bosman, SA D'economie Mixte Sportive de L'union Sportive Du Littoral de Dunkerque, Union Royale Belge des Societes de Football Association Asbl, Union des Associations Europeennes de Football; Union des Associations Europeennes de Football v. Jean-Marc Bosman. Article 177 Reference by the Cour D'appel, Liege, on the Interpretation of Article 48, 85 and 86 EC. Judgment of the European Court of Justice of 15 December 1995.' 33 *CML Rev* 991–1033.

—— and PAUL BEAUMONT (1999) EU Law (Third Edition London: Penguin).

——(2005) 'Anti-Doping Rules and EC Law' 26 *ECLR* 416–421.

WEBB, THOMAS R (1983) 'Fixing the Price Fixing Confusion: A Rule of Reason Approach' 92 *Yale LJ* 706–730.

WEILER, J H H (1999) 'The Transformation of Europe' in *The Constitution of Europe: "Do the New Clothes Have an Emperor?" And Other Essays on European Integration* (Cambridge: Cambridge University Press) 10–101.

——(2001) 'Federalism without Constitutionalism: Europe's Sonderweg' in Nicolaïdis, Kalypso and Robert Howse (eds) *The Federal Vision: Legitimacy and Levels of Governance in the United States and the European Union* (Oxford: Oxford University Press) 54–70.

WEISSKOPF, WALTER A (1965) 'Economic Growth Versus Existential Balance' 75 *Ethics* 77–86.

WERDEN, GREGORY J (1996) 'A Robust Test for Consumer Welfare Enhancing Mergers among Sellers of Differentiated Products' 44 *J Ind Econ* 409–414.

——(1998) 'Demand Elasticities in Antitrust Analysis' 66 *Antitrust LJ* 363–414.

WESSELING, REIN (1999) 'The Commission White Paper on Modernisation of EC Antitrust Law: Unspoken Consequences and Incomplete Treatment of Alternative Options' 20 *ECLR* 420–433.

——(2000) *The Modernisation of EC Antitrust Law* (Oxford: Hart).

——(2001) 'The Draft-Regulation Modernising the Competition Rules: The Commission Is Married to One Idea' 26 *ELRev* 357–378.

——(2004) 'Joined Cases C-238/99 P, C-244/99 P, C-245/99 P, C-247/99 P, C-250/99 P to C-252/99 P and C-254/99 P, Limburgse Vinyl Maatschappij NV (LVM) and Others v. Commission, [2002] ECR I-8375' 41 *CML Rev* 1141–1155.

WESSELY, THOMAS (2001) 'Case C-49/92p, Commission v. Anic, [1999] ECR I-4125; Case C-199/92p, Huls v. Commission, [1999] ECR I-4287; Case C-235/92p, Montecatini v. Commission, [1999] ECR I-4539; Judgments of 8 July 1999 (Together: Polypropylene Appeal Cases)' 38 *CML Rev* 739–765.

WESTFIELD, FRED M (1981) 'Vertical Integration: Does Product Price Rise or Fall?' 71 *Amer Econ Rev* 334–346.

WHISH, RICHARD (1985) *Competition Law* (London: Butterworths).

——(1993) *Competition Law* (Third Edition London: Butterworths).

——(1997) Exemption Criteria under Article 85(3)EC: A Study of the Case Law of the European Court of Justice the Court of First Instance and the Decisional Practice of the European Commission. *King's College London* Available from the House of Lords Library.

——(1998) 'Future Competition Law' in Ehlermann, Claus Dieter and Laraine L Laudati (eds) *European Competition Law Annual 1997: The Objectives of Competition Policy* (Oxford: Hart) 495–502.

——(2000a) 'National Courts and the White Paper: A Commentary' in 4, CELS Occasional Paper No. (ed) *The Modernisation of European Competition Law: The Next Ten Years* (Cambridge: University Printing Services, University Press) 74–78.

——(2000b) 'Regulation 2790/99: The Commission's "New Style" Block Exemption for Vertical Agreements' 37 *CML Rev* 887–924.

—— and BRENDA SUFRIN (2000) 'Community Competition Law: Notification and Individual Exemption—Goodbye to All That' in Hayton, David (ed) *Law's Future(S): British Legal Developments in the 21st Century* (Oxford: Hart) 135–159.

——(2001) *Competition Law* (Fourth Edition London: Butterworths).

——(2003) *Competition Law* (Fifth Edition London: Butterworths).

WHISH, RICHARD P (2000) 'Recent Developments in Community Competition Law 1998/99' 25 *ELRev* 219–246.

WIELEN, F G VAN DER (1971) 'ACF Chemiefarma NV v Commission of the European Communities. Case 41/69. Judgment of July 15, 1970' 8 *CML Rev* 86–89.

WILLIAMSON, OLIVER E (1968a) 'Economies as an Antitrust Defense: The Welfare Tradeoffs' 58 *Amer Econ Rev* 18–36.

——(1968b) 'Economies as an Antitrust Defense: Correction and Reply' 58 *Amer Econ Rev* 1372–1376.

WILLIAMSON, OLIVER E (1969a) 'Economies as an Antitrust Defense: Reply' 59 *Amer Econ Rev* 954–959.
——(1969b) 'Allocative Efficiency and the Limits of Antitrust' 59 *Amer Econ Rev* 105–118.
——(1974) 'The Economics of Antitrust: Transaction Cost Considerations' 122 *U Pa L Rev* 1439–1496.
——(1975) *Markets and Hierarchies, Analysis and Antitrust Implications: A Study in the Economics of Internal Organization* (New York: Free Press).
——(1977) 'Economies as an Antitrust Defense Revisited' 125 *U Pa L Rev* 699–735.
——(1979a) 'Transaction-Cost Economics: The Governance of Contractual Relations' 22 *J Law Econ* 233–261.
——(1979b) 'Assessing Vertical Market Restrictions: Antitrust Ramifications of the Transaction Cost Approach' 127 *U Pa L Rev* 953–993.
——(1981) 'Contact Analysis: The Transaction Cost Approach' in Burrows, Paul and Cento G Veljanovski (eds) *The Economic Approach to Law* (London: Butterworths) 39–60.
——(1987) 'Delimiting Antitrust' 76 *Geo LJ* 271–303.
WILLIMSKY, SONYA MARGARET (1997) 'The Concept(s) of Competition' 18 *ECLR* 54–57.
WILS, WOUTER P J (2000) 'The Undertaking as Subject of EC Competition Law and the Imputation of Infringements to Natural or Legal Persons' 25 *ELRev* 99–116.
——(2002a) 'EC Competition Fines: To Deter or Not to Deter' in Wils, Wouter P J (ed) *The Optimal Enforcement of EC Antitrust Law: Essays in Law & Economics* (The Hague: Kluwer Law International).
——(2002b) *The Optimal Enforcement of EC Antitrust Law: Essays in Law & Economics* (The Hague, London: Kluwer Law International).
——(2005) *Principles of European Asntitrust Enforcement* (Oxford: Hart).
WINSTEN, CHRISTOPHER, and MARGARET HALL (1961) 'The Measurement of Economies of Scale' 9 *J Ind Econ* 255–264.
WINTERSTEIN, ALEXANDER (1999) 'Nailing the Jellyfish: Social Security and Competition Law' 20 *ECLR* 324–333.
WIßMANN, TIM (2000) 'Decentralised Enforcement of EC Competition Law and the New Policy on Cartels-the Commission White Paper of 28th of April 1999' 23 *W Comp* 123–154.
WOLF, DIETER (1998) 'Competition Policy Objectives' in Ehlermann, Claus Dieter and Laraine L Laudati (eds) *European Competition Law Annual 1997: The Objectives of Competition Policy* (Oxford: Hart) 129–132.
WOOD, DIANE P (1993) 'User-Friendly Competition Law in the United States' in Slot, Piet Jan and Alison McDonnell (eds) *Procedure and Enforcement in EC and US Competition Law* (London: Sweet & Maxwell) 6–18.
WOOLF, LORD (1995) 'Droit Public—English Style' *Public Law* 57.
WRIGHT, VINCENT, ed. (1994) *Privatization in Western Europe: Pressures, Problems and Paradoxes* (London: Pinter).

—— and LUISA PERROTTI, eds. (2000) *Privatization and Public Policy* (Cheltenham: Edward Elgar).

YARROW, GEORGE (2002) 'Competition Policy: Its Purposes and Scope' 22 *Economic Affairs* 2–4.

YEUNG, KAREN (2004) *Securing Compliance with Competition Law: A Principled Approach* (Oxford: Hart).

ZAFIROVSKI, MILAN (2000) 'The Rational Choice Generalization of Neoclassical Economics Reconsidered: Any Theoretical Legitimation for Economic Imperialism?' 18 *Sociological Theory* 448–471.

ZERRILLO, PHILIP C, JON M FLEMMING, and ANGELA MCKEE (1997) 'Vertical Territory and Customer Resale Restrictions: A New Rule of Reason Approach' 22 *J Corp L* 705–721.

ZULEEG, MANFRED (1999) 'Enforcement of Community Law: Administrative and Criminal Sanctions in a European Setting' in Vervaele, J A E (ed) *Compliance and Enforcement of European Community Law* (The Hague: Kluwer Law International) 349–360.

ZWEIG, KONRAD (1980) *The Origins of the German Social Market Economy: The Leading Ideas and Their Intellectual Roots* (London: Adam Smith Institute).

INDEX

W

X

Y

Z

Lightning Source UK Ltd.
Milton Keynes UK
15 December 2009

147551UK00001B/67/A

9 780199 278169